U.S. Military Communications

A C³I Force Multiplier

ELECTRICAL ENGINEERING
COMMUNICATIONS AND SIGNAL PROCESSING

Raymond L. Pickholtz, Series Editor

Anton Meijer and Paul Peeters
Computer Network Architectures

Marvin K. Simon, Jim K. Omura, Robert A. Scholtz, and Barry K. Levitt
Spread Spectrum Communications, Volume I

Spread Spectrum Communications, Volume II

Spread Spectrum Communications, Volume III

William W Wu
Elements of Digital Satellite Communication: System Alternatives, Analyses, and Optimization, Volume I

Elements of Digital Satellite Communication: Channel Coding and Integrated Services Digital Satellite Networks, Volume II

Also of interest:

Victor B. Lawrence, Joseph L. Lo Cicero, and Laurence B. Milstein, editors
IEEE Communication Society's Tutorials in Modern Communications

Wushow Chou, Editor-in-Chief
Journal of Telecommunication Networks

U.S. Military Communications

A C^{3}I Force Multiplier

Fred J. Ricci
RAMCOR, Inc.

Daniel Schutzer
Citibank

COMPUTER SCIENCE PRESS

Computer Science Press
1803 Research Boulevard
Rockville, Maryland 20850
1 2 3 4 5 6   Printing   Year   90 89 88 87 86

Library of Congress Cataloging in Publication Data

Ricci, Ferdinand J., 1939-
  Military communications.
  1. Communications, Military. 2. United States—Armed Forces—Communication systems. 3. North Atlantic Treaty Organization—Armed Forces—Communication systems. I. Schutzer, Daniel, 1940-   . II. Title.
UA940.R53   1986   623.7'3   84-23783
ISBN 0-88175-016-6

CONTENTS

PREFACE

Properly understood and effectively used, military communications can provide major improvement in the deterrent posture of the United States.

The fundamental premise of *U.S. Military Communications* is that military communications systems must be robust enough to perform effectively during peacetime, crises, and hostile situations. Such capability can be achieved by a worldwide communications system that has sufficient survivability and endurability to meet the requirements of the military in extremely stressful situations. Furthermore, military communications systems must provide a "force multiplying" effect to help overcome the much larger forces of the adversary by allowing friendly units to operate in a more coordinated manner than those of their enemies.

Effective communications systems must be able to react faster and provide greater reliability, survivability, endurability, and interoperability. In addition, they must be able to control themselves so as to reduce vulnerability and, in a sense, be self-healing. Unfortunately, as noble as these goals are, they require a great deal of time to develop and large expenditures of resources to fulfill. Moreover, the physical act of communicating provides the enemy with valuable intelligence about the deployment and movement of friendly forces. These factors comprise the "Achilles heel"* in any military communication system.

U.S. Military Communications has been structured around the theme that state-of-the-art communications systems can provide the force-multiplying effect necessary for U.S. forces to be a vibrant, fast-reacting, effective military power and fulfill the command, control, communications, and intelligence (C^3I) role. Figure 1 illustrates the essential interaction of the key ingredients required to fulfill these objectives. The numbers correspond to chapter numbers. Those ingredients, listed below, are examined at length in this book.

- Requirements, operational and performance

---

*Achilles heel—a tender part of the body, easily broken but not easy to heal. The term derives from Homer's *Odyssey*, wherein Achilles, a mighty warrior, finally gets subdued by an arrow that goes into his heel.

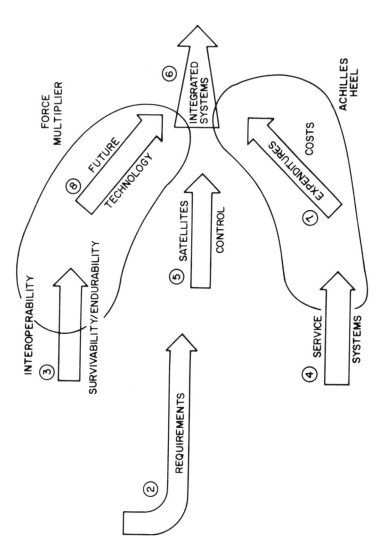

Figure 1 U.S. military communications systems structure.

- Interoperability
- Present and future service systems architectures
- New systems (e.g., satellites and control)
- Integrated systems
- Military expenditures
- Future technology

The thrust of the examination is to explore the stringent and often conflicting survivability, performance, error rate, capacity requirements of security, anti-jamming, speed of service, and the interplay of interoperability and survivability, showing how these elements affect the design and development of future systems.

Chapter 1 provides an introduction that lays the framework for the remainder of the book. Chapter 2 discusses the unique communications requirements of the military. This includes the stringent requirements of mobility, jamming protection, and survivability. The discussion includes the development of a generic model and the presentation of specific examples.

Chapter 3 summarizes the key ingredients of U.S. and NATO systems and examines interoperability issues in detail.

Chapter 4 provides a framework for understanding the systems architecture of the respective services, indicating which systems are available and which are planned for the future, and discussing architecturally related considerations.

Chapter 5 focuses specifically on satellites and control, because of the significant role that the latest survivable satellite systems and control will play in producing the force-multiplying effect.

Chapter 6 looks at an integrated systems approach utilizing artificial intelligence (AI) techniques, indicating how information flow can be optimized to provide a force-multiplying effect that depends on future technology and expenditures.

Chapter 7 examines techniques for selecting and developing systems and considers expenditures and investment strategies for the defense industrial base.

Chapter 8 explores the potential of several key future technologies, and assesses their impact on telecommunications.

Clearly, a force-multiplying effect can be achieved if future military communications systems are designed to take advantage of many of these technological advances and if these designs can be fielded in a timely fashion. Without an effective communications capability, the missions and functions of C^3I cannot be fulfilled.

U.S. Military Communications provides a comprehensive ''systems'' view of military communications systems that will stimulate readers and answer some critical questions. In addition, some new perspectives, including relationships to artificial intelligence, warfare strategies, and security are presented. Some important aspects and quantitative data have not been included, because of their classified status; all technical and programmatic information provided here has been derived from unclassified sources.

The design, development, and use of military communications systems is unique because of the need for survivability, resilience to countermeasures and electronic warfare, radio electronic combat, antijamming for digital data and voice, and communications with reconnaissance systems and smart weapons.

Many exciting changes are taking place in military and commercial communications as a result of advances in digital switching, networking, signal processing, very large scale integration, very high speed integrated circuits, integrated microwave technology, microcomputers, millimeter waves, fiber optics, satellites, and laser technologies. *U.S. Military Communications* addresses the latest advances in those technologies and considers how they are being utilized in military systems to meet the challenges of the 1980s and 90s. Both strategic and tactical military systems are described. The present and future Defense Communication System (DCS) and NATO systems are discussed in the context of how they interconnect and interoperate with each other and with other major systems. Army, Navy, Air Force, and Marine Corps tactical communications related to surveillance, control, and force coordination are treated in detail. The need for priority, preemption, survivability, antijam, low probability of intercept, and real-time responsiveness and its impact on the design of communication systems is presented along with specific examples.

The perspective taken in *U.S. Military Communications* is that operational requirements determine the ultimate design, development, and fielding of military communications systems. Therefore, to fully appreciate the design objectives for a military communication system, one must first understand the underlying operational requirements it is intended to satisfy. Since military requirements differ from commercial requirements in several fundamental respects (e.g., survivability, resistance to countermeasures, operating environment), those communications systems are not just militarized mirror images of commercial systems.

State-of-the-art military systems have been documented so as to provide a comprehensive treatment of strategic and tactical systems that satisfy critical, stressful situations. In addition, the systems aspects of military communications are considered from the perspective of "information" management and artificial intelligence. After all, the real interest of the communications network is to provide a mechanism for accurate, reliable, information flow that helps to manage one's own forces and to detect, target, and attack the enemy in such a fashion as to prevent or win battles. Strategic or tactical aspects of information flow, intelligence-gathering systems, and command centers are considered since communications systems cannot be viewed in isolation if an effective war-fighting capability is to be achieved.

At present, there is very little information documenting military communications systems. In 1980 a special issue of the *IEEE Transactions on Communications* was devoted to military systems, but there is a minimum amount of information in textbook form. *U.S. Military Communications* fills a void in the literature that has existed for some time. It is written for the practicing engineer and for military personnel, as well as for the student. In many applications, the development of a military communications system stresses the state-of-the-art in technology. In fact, communications techniques and concepts under development in the military are often precursors of developments that ultimately become important commercial systems. Therefore, practicing engineers working in commercial areas, as well as those in the defense portion of the industry, will find this book valuable. Students at both the undergraduate and graduate level will find it to be a comprehensive text and reference guide.

The views and judgments expressed in *U.S. Military Communications* are solely those of the authors and do not necessarily reflect endorsement by their respective employers, the U.S. government, or the publisher.

Fred J. Ricci
Daniel Schutzer

ACKNOWLEDGMENTS

The writing of a book of this kind involves the help, advice, and encouragement of many people. Although we have enjoyed developing the ideas, concepts, and technical material for this volume, the tediousness of the work had to be offset by the support and humor of our families and friends. Many thanks to Myra and Mary Jo, our beloved wives, and our children, Ferdinand, Danté, Eric, Richard, and Pamela.

Special thanks are due Dr. Raymond Pickholtz for his encouragement, advice, and helpful review. The contributions of Stephen B. Heppe, of Stanford Telecommunications, Inc., in the area of satellite control are greatly appreciated. Clark Edwards' contributions on strategic/tactical interfaces also lent a new look at a difficult problem. Jerome Blackman (deceased) of the Army Communications-Electronics Command (CECOM) contributed to the discussion of Army doctrine. Dr. Fred Ellersick of the MITRE Corporation has contributed information on Air Force communications. Dr. Edwin L. Woisard of RAMCOR, Inc., contributed his expertise on military operations and communications and reviewed this work for editorial corrections. Drs. Neil Birch and Richard DeNucci have contributed by providing helpful suggestions and editing. Dr. William Rakestraw has contributed a great deal by being a final reviewer and sharing information on digital systems, electronic warfare, and advanced technology. Dr. Jon Boyes and K. Lamar of the Armed Forces Communications and Electronics Association (AFCEA) have also provided many helpful suggestions and the benefit of their review.

Chapter 1

INTRODUCTION

It has been said that the best deterrent to war is a strong, flexible, quick-reacting military supported by a cornerstone of telecommunications.

America's defense rests on the following: the ability to obtain strategic and tactical intelligence; to effectively control its resources over a global battlefield; to permit instantaneous communications with its operational forces; to simultaneously identify and track a multitude of targets under and on the sea, on the ground, in the air, or in space; to utilize and direct appropriate weapons against these targets under any conditions, at any time, with pinpoint accuracy; and finally to subvert the effort of opposing forces to do the same. Among the means for accomplishing these tasks are electronic systems that sense, transmit, compile, analyze, display, compare, store, and process information and then launch, guide, control, and trigger the appropriate weapons systems.

Both the Falkland Islands situation and conflicts in the Middle East underscore the fact that electronics has become the keystone to a credible defense. The U.S. Department of Defense (DoD) recently stated that "Electronics is the most critical of all technologies for the maintenance of peace." [1] Without question, the unparalleled strength of U.S. defense capabilities depends absolutely on electronics in one form or another. Moreover, "electronics is the *force multiplier*." [1] It permits the United States to deter aggression without imposing a burden that would inhibit continued improvement in the quality of life for its people and those of its allies.

Among the most prevalent of military electronics systems is communications. Indeed, communications can be said to provide the glue that holds the rest of the system together. This book considers the state of the art in military communications systems for both tactical and strategic use in terms of their role in supporting command, control, communications, and intelligence (C^3I) military operations. The intent is to discuss both the *practical* and *theoretical* aspects of communications related to meeting the most up-to-date requirements of land, sea, and air warfare situations. Chapter 2 begins an exploration of these requirements.

The subject of military communications is vast and complex, especially when all of the services of the United States and its allied (e.g., NATO) forces must operate together in the same theater of operations. Figure 1.1 depicts the extent of communications systems considered within this book for U.S. Army, Navy, Air Force, Marine Corps, and NASA usage in land, air, and sea battle situations. Examples of current military communications systems are provided in Chapter 4. Figure 1.2 depicts scenarios in which military communications systems will be utilized in the future ranging from extremely low frequency (ELF) to extremely high frequency (EHF).

The military communications systems planned for the future are intended to supplement current systems by providing high-capacity, secure, jam-resistant (AJ), low-probability-of-interception (LPI), electromagnetic pulse (EMP) pro-

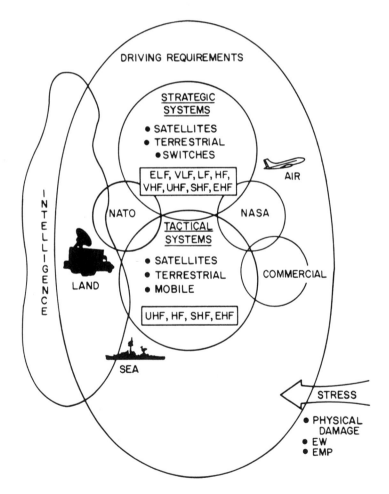

Figure 1.1 Overview of U.S. military communications systems in a C³I
environment.

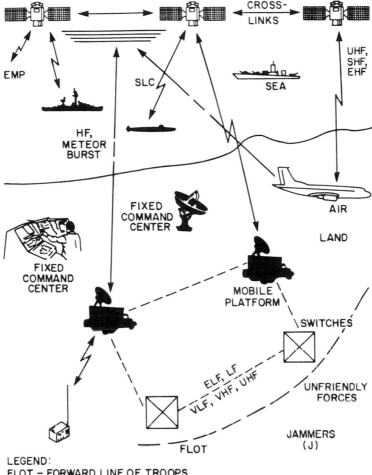

Figure 1.2 Generic view of future military communications systems.

tected capabilities that meet the stringent requirements of a fast-moving enduring battlefield situation.

The use of satellites and mobile communications assets to command and control strategic and tactical use of land, sea, and air forces will increasingly be a key ingredient of operations in the future. Fixed-plant terrestrial systems will play less of a role. Transportable tactical systems with secure voice, data, and facsimile capability will provide communications for land forces. Mobile multichannel jam-resistan. communications will replace single-channel VHF radios for use by the foot soldier. Important interfaces with NATO Public Telephone and Telegraph (PT&T), NASA, and commercial communications will require increased emphasis on interoperability and standards. Why is this so? Chapters 6, 7, and 8 discuss future trends in warfare and communications electronic technology. The role of communications in supporting these trends is explored in some detail.

Although not much thought is generally given to the use of commercial communications to support military operations, a great deal of the American Telephone and Telegraph (AT&T) switching and transmission facilities, in addition to NASA and NATO resources, are available for use in the future for military purposes. For example, the worldwide resources of the NASA tracking network could be utilized for military purposes. Commercial satellites may also play a major role. In general, new approaches to utilizing existing U.S. communications resources to meet changing military requirements must be found in the future to provide the quick response, flexibility, and surge capacity needed in wartime situations. With the advent of the Air Force, Navy, and Army Space commands, elements such as the Consolidated Space Operations Center (CSOC) are playing greater roles in carrying out strategic surveillance, command, control, and communications missions.

Hopefully, within the constraints allowed in a book of this kind, the reader will gain an appreciation of the important issues of present and future military communications systems. There are many popular topics that have not been treated extensively—high frequency networking, packet switching, packet radio, VHSIC processing, spread-spectrum processing, etc. This is because there is simply not enough space to treat every topic in detail. The proceedings of the IEEE Military Communications Conference (MILCOM) should serve to provide the latest, state-of-the-art, unclassified, and classified information to practicing engineers. Reports from such agencies as the Defense Communications Agency (DCA), the National Aeronautical and Space Administration (NASA), the Government Services Agency (GSA), and the various military services should serve as excellent sources of material on technological advances.

1.1 REFERENCE

1. *IEEE Spectrum on Technology in War and Peace* (October 1982).

Chapter 2

REQUIREMENTS FOR MILITARY COMMUNICATIONS

2.1 FUTURE DIRECTIONS

At this point it is useful to review the evolving role of communications in the military. How are the military's communications needs changing and where are today's communications-related technology trends likely to lead us? First, it is important to realize that military communication is not an end in itself. The military should communicate only when it is absolutely required, when communication is an essential step to accomplishing some military objective, aside from routine traffic. This is because, in general, military communication always has associated with it some attendant risks and vulnerabilities. Thus, communication in support of some operation is to be avoided wherever possible.

Why is it that military communications represent such a risk to military operation? In order to communicate over any appreciable distance, information must be transformed to some form of energy fluctuation (usually electromagnetic) that can be transported to and detected and measured at the receiving end. This means that at any given instant in time, these energy fluctuations must be present in some finite volume in space and time that lies somewhere on a path or paths between the transmitter and the intended receiver(s). They may also propagate along other paths that do not include the intended receiver. In principle, any receiver that is suitably designed can recognize and detect these fluctuations as long as it is placed along one of these paths at the right time. If this receiver belongs to the enemy, he may use the received communications intelligence to his advantage. At a minimum, he is alerted to the presence of a transmitting party somewhere along the transmission path. In some cases, the location of the detected transmitter can be isolated to a likely region along the transmission path. In still other cases, the identification of the transmitter, and sometimes the receiver, can be determined. Sometimes the purpose and even the contents of the communications can be determined.

Knowledge of any of this information puts the intercepting party at an advantage that can be exploited to anticipate and/or react in a timely manner to prevent or to counter the intended military operation that the intercepted communications are supporting. Further, if a transmitter, such as a jammer, or some uncontrollable environmental disturbance can be placed along the transmission path, then the desired transmission can be degraded, disrupted, delayed, or in other ways interfered with, including the possibility of the insertion of false data (spoofing).

Consequently, as long as the communications transmission paths include spaces that are not entirely under the exclusive control of friendly forces, when a military unit communicates, it communicates at risk of interception, exploitation, degradation, and deception that can compromise and/or disrupt the military operation this communication is intended to support. In military operations, the more restrictive and local the volume that the communications paths can be confined to in space and time, the smaller will be the risk associated with the communications.

However, it is clear that the larger the separation between the parties that wish to communicate and the greater the number of parties that wish to communicate, as well as the greater the volume of information that needs to be exchanged among the communicating parties, the more difficult will be the job of restricting the space and time associated with a communications transmission. This is one of the Achilles heels of communications. In fact, military operations are growing in the very direction where greater volumes of information need to be exchanged among more widely dispersed combat units and greater numbers of units need to be involved in these communications. Why is this so? Certain basic objectives and principles that are inherently desirable for any military combat operation, when combined with the recent dramatic advances in military weapons and sensor technologies, serve to drive military communications requirements in these directions. The key underlying principles are summarized below:

- *Law of large numbers*. Assuming relative parity in the fighting capability of the combat units involved in an engagement, the greater the disparity in the number of units engaged in a two-sided battle the greater are the odds of victory for the side with the larger number of units. Lanchester's equations (explained in Chapter 6) dictate that this advantage can be as high as the square of the difference between the number of combatant units involved. This principle is related to the following desirable attributes/objectives.

- *Mobility*. The mobility of a unit is directly related to its ability to maneuver and to mass and concentrate (increase) the number of units that can be employed at the scene of battle to attack the enemy. Mobility also enhances a unit's survivability in terms of its ability to outrun or disperse in the face of superior enemy attack.

- *Area of influence*. The area of influence is the area within which a combat unit can inflict harm to an opposing unit. Clearly, the larger a unit's area of influence, the less maneuver is required to mass and bring to bear a large number of dispersed units. And, if there is a disparity in the relative areas of influence of two opposing units, then the unit with the larger area of influence will generally be able to strike first with impunity while still out of range of the opposing force's ability to launch a counteroffensive.

- *Area of perception.* The area of perception is the area within which a combat unit can sense and keep aware of the activity and movement of forces. This is generally, or should be, larger than the unit's area of influence. It allows the planning time necessary to make decisions with respect to the proper allocation and exercise of the combat unit's weapons to the full range of its area of influence.

- *Surprise.* The element of surprise relates to the extent to which a combat unit can take action without tipoff of its intentions to the opposing force. The further into the battle plane that the combat unit can get prior to discovery, the less is the available reaction time of the opposing force, the fewer are the opposing force's options, and the greater is its confusion. This surprise is achieved through the orchestration of a combination of elements such as increased speed of operation, synchronization of the actions of dispersed units over the entire operation, and proper integration with coordinated camouflage, cover, and deception (C&D) actions (covert operation, jamming, decoying). This implies a great deal of prior planning as well as coordination throughout the combat operation. It also implies a speeding up of the tempo of combat operations.

- *Tempo.* Tempo is the rate of change associated with the combat environment. The combination of increasing a combat unit's mobility, area of influence, and area of perception, coupled with a desire to maximize the element of surprise in a combat operation, dictates an increase in the tempo of combat from planning to execution.

Today's technology permits us to improve combat in almost all of these areas. Extended weapons ranges and increasing speeds expand a unit's area of influence and the tempo of operations. More sensitive sensors, and satellites, and longer endurance aircraft extend the combat unit's "eyes" or area of perception (combat horizon). Advances in computer technology and software shorten the processing time for detection, classification, identification, and decision making, allowing an increase in the tempo. Even though the absolute number of units may not be growing as dramatically as these other areas, the increased reliability, mobility, and extended range of influence of modern combat units make the number of units available at any time to support a given combat situation much higher than previously. In short, the forces that need to be coordinated are increasing in number, are getting more dispersed, and require a greater volume of coordination and information exchange over shorter periods of time.

Automation and communications support the growing requirement for rapid decision making. In battle, the Army, Air Force, Navy, and Marines will need the full application of automation/communications capabilities to achieve their mission. Clearly, the management and proper exploitation of information is a force multiplier. The increasing complexity, speed, and lethality of modern warfare dictate that commanders and their staffs at all echelons be able to make timely, accurate, and effective decisions. The time available for collecting and evaluating information, making decisions, and disseminating directives is being compressed both by the capabilities of other nations and the increasing complexity of our emerging weapons system technology.

As a result of the planned increase in automated systems for the battlefield, there will be a subsequent increase in the requirement for data distribution for most of these systems. For example, in the near future, a number of battlefield automated systems will be procured to provide automation for intelligence, logistics, administration, air defense, field artillery, communications, and command and control. In order for these systems to perform their designated functions and interface with one another, the associated telecommunications equipment must also be procured.

As time goes on, an increasing number of automated systems will be fielded. It is estimated that there will be a 25 percent increase in the amount of communications needed for data on the battlefield and that need will increase exponentially over the next five years. Eventually, the battlefield will reflect a proliferation of automated systems. In short, communications requirements are getting harder to meet, and unless these more stringent communications objectives can be met reliably and securely, they will become the weak link in any combat operation.

What techniques and future technological improvements can be applied to achieve these objectives, to allow units to communicate more securely (reduce the space-time volume over which a communications transmission can be interrupted or interfered with by a uninvited third party) while supporting increased communications volumes among an increased number of units dispersed over a larger geographic area more reliably? There are several approaches, technologies, and trends:

- Higher frequencies (more narrow beamwidth, greater potential capacity—an exception is the trend back to HF)
- Short signals, spread spectrum (including frequency hopping) and multicarriers, orthogonal coding
- Packet routing and other distributed computer networking techniques coupled with greater line-of-sight (LOS) relaying
- Local area networks (LANs)
- Power control
- Adaptive antenna nulling
- Fiber optic technology
- Artificial intelligence and natural language

These approaches and technologies are discussed in more detail in Chapter 8, as well as the problem of how to address, acknowledge, and dynamically task and reallocate functions between highly mobile units that like to be covert and silent most of the time and where the number and types of units involved can change over time (units get born and die). A unified framework for reviewing future communication systems design in light of these trends and requirements is presented in Chapter 6.

2.1.1 The Role of Electronic Warfare

The importance of electronic warfare (EW) has grown significantly in the last few years. The need for both offensive and defensive combat will increase as we progress toward the concepts of Air-Land Battle 2000. Prior to describing specific equipment and capabilities, it is important to understand the role of EW in the battle.

The battlefield of the future has the following characteristics:

- Large quantities of sophisticated combat systems
- Difficult command/control
- No significant qualitative advantage
- Battle expanded deep into enemy air space and formations
- Intensive battle at the decisive points

The salient feature of the concept is *maneuver*, which enables a command to place the enemy in a position of disadvantage. We must know what the enemy is doing (both up front and deep); we must be able to acquire, target, and destroy key elements; we must maneuver and allow our forces the staying power to destroy the enemy; we must be able to make quick, accurate, and timely decisions; and we must be able to synchronize the options of the combined and supporting arms.

When properly employed, electronic warfare can be expected to make significant contributions. The primary payoff is increased combat effectiveness through real-time assessment of the hostile battlefield environment, quicker and more accurate counterfire and electronic countermeasures (ECM), increased survivability of personnel and equipment, and sustained communications, command and control operations (C^3) despite hostile EW actions.

The role of EW in this complicated situation can be seen by examining the elements of a combat situation.

2.1.1.1 *Target Servicing and Suppression/Counterfire*

Target servicing systems must be able to operate in an intense ECM and air defense environment. Suppression/counterfire systems must be able to locate, jam, and/or destroy the enemy weapon systems. The role of EW in both these areas is to:

- Improve survivability of suppression/counterfire systems
- Improve survivability of air and ground combat vehicles

- Locate and classify critical communications/noncommunications targets
- Jam enemy communications/noncommunications systems
- Degrade the enemy's ability to see by degrading electrooptical systems
- Harden one's own systems to EW

2.1.1.2   Air Defense

In face of fixed-wing and helicopter aircraft presence, an effective air defense is required and must operate in an intense ECM environment. The role of EW is to:

- Reduce vulnerability to enemy EW
- Detect, identify, and jam enemy terrain-following radars
- Detect, identify, and jam enemy avionics signals
- Protect against antiradiation and IR/EO missiles (C^3 EW)

C^3 EW refers to the U.S. ability to operate its communications, command, and control systems in a highly sophisticated enemy EW environment. The role of EW here is to:

- Provide EW training and test devices to determine C^3 vulnerability
- Assist in the development/evaluation of hardened communications systems
- Develop electronic counter-countermeasures (ECCM) to negate the effectiveness of enemy jamming

2.1.1.3   Tactical SIGINT

Signals intelligence (SIGINT) is the ability to "see" the enemy both up front and deep by collecting, processing, and analyzing enemy communications and noncommunication electromagnetic signals. Thus, SIGINT is generally divided into two categories: COMINT (communications intelligence) and ELINT (electronic, or noncommunications, intelligence).

COMINT is the collection and analysis of data on an enemy's communication. ELINT is the collection and analysis of data on signals other than communications, such as radar.

2.1.2 Systems Design

The communications engineer must be aware of the hostile EW threat and the underlying signal structure and must design systems so that communication is not vulnerable to jamming and exploitation. The key interests of the communications engineer are to:

- Use EW and tactical SIGINT as a force multiplier to reduce an enemy's numerical superiority

- Provide immediate answers to battlefield commanders
- See deep into enemy territory to give early warning of intentions and capabilities
- Increase survivability, and therefore mission effectiveness, of aircraft and ground combat vehicles
- Protect C^3 systems from enemy EW and SIGINT
- Produce systems that are reliable and easy to use and maintain

It is clear that the driving forces behind all product improvements are the threat of conflict itself and the needs of the battlefield commander in the face of dollar, time, and mobility constraints. The United States must be prepared to handle a dynamic, complex environment that is continuously evolving. It must be attuned to the latest technological developments. Future SIGINT/EW equipment will incorporate modular concepts, make extensive use of adaptive processing/artificial intelligence, and be designed for technology insertion. These characteristics will ensure that all systems deal successfully with emerging hostile emitter environments in a way that is responsive to a commander's needs.

The critical technical challenges to achieve and maintain this capability are to:

- Handle an increasingly dense environment
- Field threat-reactive systems
- Extend range and frequency coverage
- Develop systems to locate, identify, and jam exotic modulations
- Improve classification/identification of targets or threats
- Improve targeting accuracy
- Speed targeting data to fire elements
- Develop countermeasures against electrooptical threats
- Decrease vulnerability to enemy SIGINT/EW

The Yom Kippur War (1973) was a practical example of the use of Soviet EW equipment to intercept, jam, and/or exploit communications systems. In the face of the ability to interrupt communications, the challenge for the United States is to provide a force multiplier by designing communications systems to overcome these threats. Section 2.5 gives a specific example of the design of a tactical communications system considering aspects of jamming.

Other factors affecting the design of communications systems for military applications include the side effects of nuclear warfare:

- EMP (electromagnetic pulse)
- SGEMP (system-generated EMP)
- Radiation effects (α, β, γ)
- Dust clouds

- Ionization effects
- Blast

A considerable amount of work has been done to overcome these effects by industry, the Defense Nuclear Agency (DNA), and others. Most of the aspects of these effects are classified and cannot be considered in this book.

The communications engineer should be aware of these effects and take them into consideration when designing systems. For example, the transponders aboard a satellite must not only provide for jam resistance but also be shielded against radiation effects. Ground-based systems must be hardened to protect them from EMP.

2.2 INTERFACES AND SERVICES

The purpose of this section is to establish a foundation upon which secure C^3I information systems may retain cohesive connectivity in the face of battle.

Command centers, both those of a purely civilian nature as well as those designed to carry on military operations, are primarily information systems: intelligence, reconnaissance, weather, logistics, and unit status information flows into them, while decisions concerning unit deployment, weapons utilization, and additional intelligence collection needs flow from them. Whereas voice communications support a great deal of the minute-by-minute directing of tactical operations, authenticated record messages are used as data sources for large-scale, computer-based information processing and display systems. Command centers are important to the conduct of battle, of course, but without the switching and transmission systems that interconnect them and move information from one point to another among the various information sources and sinks, these centers can become the foci of disinformation and confusion, and decision-making uncertainty.

During the various stages of conflict, from peace through crisis, through limited, general, and nuclear war, there is a transition of functions from the highly central strategic chain of command to the more local tactical chain of command. This transfer must be supported by continuous secure communications connectivity. The current system of switching centers and transmission systems evolved from the early days of telephone switching centers and transmission systems without any regard for system security or service continuity during the spectrum of adverse conditions characterized by the range of hostilities encountered in modern warfare. These facilities came into being as the demand for service exceeded the physical capacity of wire routes and as mechanical and electronic advances made modern network switch designs a practical reality. Later, as transmission line theory and multiplex concepts merged and were applied to telecommunications, the facilities became more complex physically, but the underlying theory of operation and motivation changed very little. Security and survivability measures, when added, were an afterthought. The transition from a fundamentally nonsecure peacetime communications system to a secure survivable system design must evolve from the current plain-text/cipher-text transition point outward, in both directions simultaneously, and thus must rely on control theory as well as information theory to achieve a workable, end-to-end secure and survivable system.

In Figure 2.1, [1] the fundamental network design based on switches and transmission media is clearly evident. Of critical importance, however, in establishing a secure network concept that can withstand disruption are two elements upon which the need for networks is based. The first element is that of chain of command; if it is assumed that central direction and control of a fighting force is essential to successful conduct of war, then not only must intermediate command elements be prepared to be bypassed when the communications elements serving them are disrupted, but the central or focal point of command must be prepared to cope with large quantities of information flowing into it without benefit of the human filtering process that takes place at intermediate points along the way [2]. Similarly, an increase in outbound traffic may be expected because of the need to connect directly to many low-level units formerly subordinated to a few mid-echelon command activities.

The second factor of network needs is related to the first. Once a chain of command has been established and the corresponding hierarchy of telecommunications networks has been brought into physical being, the human element of information transfer imposes itself upon the electronic networks thus created. Command and control begins to take place and the functional subnetworks of intelligence, operations, logistics, plans, and communications control, etc., are developed utilizing voice and formatted record messages to support the management concept of military hierarchies. Unfortunately, once in place and functioning, the telecommunications facilities serving each subnetwork at its functional level of command are seen by the other functionaries within the respective command elements as nonresponsive to their needs. Consequently, for all but administrative purposes, excluding voice networks, the formatted record message systems proliferate and take on many formats or protocols each in support of another functional command element. The result is a mix of special subnetworks and formats that differ greatly, especially between strategic and tactical telecommunications systems.

Thus in the design of new secure telecommunications networks and physical components, unless the format interfaces between the old and new components are accounted for, the system will not be able to be easily interfaced and reconstructed to withstand physical destruction and electronic interference. With the system unable to reassemble a functioning telecommunications network to maintain some semblance of chain of command, tactical commanders must be prepared to function on their own and central command will need to consider the ways and means to cope with the lack of connectivity to the tactical element and with misinformation regarding situations ranging from crisis to broader strategic warfare.

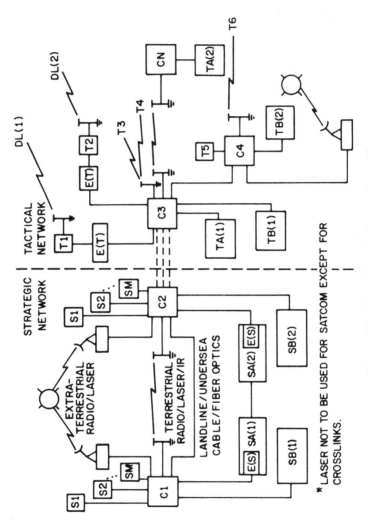

Figure 2.1   Strategic/tactical network model.

The need for configuration management on a large-scale network basis is also evident, particularly when decisions are required to validate and initiate new system acquisition programs.

How then may a cohesive interconnect design evolve from such anarchy? The various internal elements of the switching and transmission systems of the secure communications network model in Figure 2.1 can be broken down into standardized interfaces. Detailed design specifications document those interfaces that may account for the hierarchical military chain of command and the functional elements concerned with specific formatted message exchanges [3]. They point out the key interface points and functions where security must be designed in from the beginning.

Secure voice operations can be made interoperable through a variation of the concept of a standardized numbering plan and through some standardization of the voice digitization schemes. The more difficult case of interface definition in light of network disruption is that of secure, formatted record message transfers, because of the existence of disparate data element standards.

Using as a reference the symbology in Table 2.1, the following description of the effect of physical disruption within a network indicates the nature of the interface problem and points toward a means of resolving the difficulties postulated. Assuming disruption at tactical switch TA(1) between a tactical pro-

Table 2.1 Table of symbology

Strategic network	
S1, . . ., SN, SM	—Strategic subscribers (voice, data, facsimile, etc.)
SA(1), SA(2), . . ., SA(N)	—Formatted messages switches (DDN, SACDIN, etc.)
SB(1), SB(2), . . ., SB(N)	—Dedicated voice switches (FTS, DSN, etc.)
E(S)	—Encryption system(s) for strategic networks
DDN	—Defense data network
DSN	—Defense switched network
Tactical network	
DL(1), DL(2), . . ., DL(N)	—Data links (JTIDS, SINCGARS, PLRS, etc.) Note: Each data link may be encrypted using a unique keying variable.
T1, T2, . . ., T6, . . .	—Tactical subscribers (voice, data, facsimile, etc.) (highly mobile)
TA(1), TA(2), . . .	—Tactical message switches (formatted record traffic)
TB(1), TB(2), . . .	—Tactical voice switches (TRI-TAC, Airborne C3, etc.)
E(T)	—Encryption system(s) for tactical networks
Control element	
C1, C2, C3, . . ., CN	—Communications Technical Control (DCAOC, DSCS Control Element, TRI-TAC CNCE, etc.)

cessing center T1 and a strategic command element subscriber at S1 served by strategic switch SA(1), several alternate paths exist to effect the transfer of formatted messages from T1 to S1 and vice versa. If the tactical control center C3 remains connected to the strategic control center C2, and C2 to C1, messages encrypted in keying variable E(T) may be manually routed to the strategic switch SA(1) serving subscriber S1. The distribution of clear text messages to S1 presumes, though, that switch SA(1) is equipped not only to decrypt messages encrypted in E(T), but also to store both the necessary routing and control information to acknowledge traffic from center T1 and routine information and to relay what may be defined by T1 as high precedence information. If, in addition, the information generated at T1 was previously routed to another strategic subscriber S3 and S3 prepared summary information based on inputs from tactical sources T3 and T6 for subsequent distribution to S1, then a determination as to the urgency of the summary information is required so as not to overload switch SA(1) when direct connection to T1 is effected.

Thus, a new interface has developed unexpectedly and may be dealt with by documenting, and placing under controlled distribution, the expected interfaces. Several factors must be considered, including security (keying variables E(T) and E(S)), communications control procedures, and standard data control protocols (including programming languages, route control formats, storage mechanisms, and a standard data element structure for the communications processor function). In the to/from sense, the compatibility of control elements C1, C2, and C3 is essential down to and including telecommunications security devices and the voice or teletype terminal devices. Circuit conditioning capability must have the same performance parameters.

At the switching level, normally connected message switches SA(2) and TA(1) must have cryptographic equipment and keying variable compatibility, the same route tables, the same formatting capability, and must also include contingency switching ability in either direction. Voice switches SB(2) and TB(2) must also be cryptographically compatible, contain the same numbering plans, and also be of the same nature, e.g., space division and/or time division, and under the same control scheme, e.g., CCIS or CCITT No. 7.

The various factors of potential incompatibilities between the diverse worlds of strategic and tactical telecommunications networks, both containing the same generic elements of switching mechanisms and transmission media, may be accommodated through cooperation between network users, system designers, and acquisition managers, provided the interface control documentation and configuration management procedures needed to effect the dissolution of incompatibilities are forthcoming. To ensure that new information processing systems are compatible with existing networks and may thus be properly utilized by an ever-changing force structure, the preparation of a data item description (DID, in Department of Defense language) that delineates a secure communications plan and becomes a part of planning contracts and subsequent system acquisition actions appears to be necessary.

This description, based on large-scale network theory, would encompass the security, communications control, and protocol factors discussed above as well as the governing concept of operations. A document prepared as a high-level planning tool would serve as a master configuration control plan and become the controlling factor in assessing mission needs, validating those needs, and deciding upon an acquisition strategy to satisfy the defined need.

In addition, the preparation of tables, not unlike the tables of routing indicators used by military communications operations personnel, that would be used by military communications technical control personnel is needed to effect the deployment of systems [1]. For these new systems, the tables would have to be extended to include parameters associated with security, communications and data control, and connectivity to ensure rapid restoral of interconnects among systems. Table 2.2 is presented as an example.

The problems associated with continuing to independently analyze user needs, to develop new systems without benefit of an understanding of existing systems,

Table 2.2 Connectivity for station T1*

Station T1*	
Security:	Cryptographic device—data encryption standard Mode—1 Keying variable—MAT 42 Change time—0900Z Circuit designator—JBV863 Controlling authority—NMIC
Communications control:	Primary control—C3 Secondary control—C4 Orderwire terminal—AN/UGC-141 Protocol—ASCII Data rate—1200 BPS Format—From/to station ID/TOD/OP ID/TEXT/EOT
Data control protocol:	Language of processor—Fortran IV Route control format—RXXXXX Storage mechanism—Disk pack (direct access using routing indicator YRDJTF) Data element structure—NN/ROUTE/FILE/CRC/EOT Cyclic redundant code format—XXXXXXXX
Connectivity	Normal—Tactical switch TA(1) via wire pair 7X8 Alt capability—Tactical switch TA(1) via AN/TRC-170 on Multiplex Group 7, Channel 3 —Strategic switch SA(2) via C2 on Trunk Group DRJI0938 —Strategic switch SA(1) via C2 on Trunk Group LBDA2401 and Wire Pair 4K33

*Insert system/program title/station ID/exercise name/operations plan (as appropriate).

and to deploy those new systems into the hands of communications operations personnel unprepared for an increased level of complexity will likely have severe consequences on future battlefields. An organization comprised of individuals selected to perform battle management activities from afar who take for granted ideal communications connectivity conditions develops a mindset or psychology of momentum and ebb and flow of battle actions not unlike childlike hyperactivity. Loss of communications connectivity may prove both unsettling to the individuals and fatally disruptive to the organization and what it represents. Communication systems designers and network operators must be cognizant of this psychological concern, for they may be the first to experience the consequences of a failure to communicate [2].

2.3 TACTICAL COMMUNICATIONS FOR THE AIR-LAND BATTLE

The former Chief of Staff of the Army, and a strong proponent of the Army Air-Land Battle approach, caused profound changes in the tactical doctrine textbooks. As discussed in an article in the *New York Times Magazine* of November 28, 1982, entitled ''The Army's New Fighting Doctrine,'' General Meyer, using the tactical brilliance of past compaigns displayed by generals such as George S. Patton and Douglas MacArthur, directed that the Army become a lean, agile, long-legged, counterpuncher. The goal of the new doctrine would be to ''disrupt the enemy timetable, complicate his command and control, and frustrate his plan, thus weakening his grasp on the initiative.'' That philosophy has not changed. Requirements and designs for tanks, guns, aircraft, and armored vehicles are being formulated to implement the doctrine. Similar initiatives are also being carried out by the Air Force, Navy, and Marine Corps.

Significant changes in the Army's communications are needed to support this new doctrine if it is to succeed [4,5]. We can no longer afford the thousands of vehicles, personnel, and hours or days that are necessary to provide the communications for the current styles of tactical deployment and tactical doctrine. Communications must match the maneuverability of the troops. Communications must no longer hinder the mobility of the headquarters of the different echelons of command as the troops move about the battlefield. Communications must support the ability of the headquarters to be dispersed rather than centralized. The ultimate goal of communications facilities is not to be transparent, but to be invisible.

All existing communications facilities and equipment need a thorough reevaluation of their capabilities and characteristics. Switching, transmission, multiplexing, terminals, patching, network control, and security will be subject to microscopic examination. Much of the equipment in use today and some to be fielded in the near future are likely to be too heavy, too noisy, too large, consume too much power, and may not provide those capabilities required by the Air-Land Battle doctrine. Communications equipment being marketed in the commercial environment is taking advantage of the latest technologies to become lighter, quieter, cooler, and smaller. The Army intends to take advantage of the same technologies and still accommodate the need to provide security and protection against the threats of electronic countermeasures as well as the nuclear battlefield. Those personnel who must operate and use tactical communications equipment are going to appreciate fully the need to make the equipment and facilities "invisible."

With support and emphasis from the highest levels of the Department of Defense and the Department of the Army, a new attitude is being formed [6]. The U.S. Army Training and Doctrine Command, at Fort Monroe, Virginia, and the U.S. Army Material Development and Readiness Command, at Alexandria, Virginia, are cooperating in a joint venture to ensure that the appropriate communications equipment will be available when it is needed to support the new Air-Land Battle doctrine. The necessary requirements documents are being generated, coordinated, and approved in weeks rather than months or even years, as was usually the case. Parochialism is disappearing and is being replaced by a new spirit of cooperation unheard of in the not-too-distant past. This new spirit is growing in two directions: from the highest echelons of the Army downward and from the lowest levels upward. With continued emphasis and the determination of all Army personnel nurturing it, there is an excellent chance that this new spirit will grow into a long-lasting attitude.

In the winter of 1983, an intense effort was pursued to examine the existing requirements for communications equipment. The effort was called a "scrub" and one of its purposes was to attempt to remove from the requirements documents those specified functions that would inhibit the capability of the equipment to support the new doctrines. Another purpose of the "scrub" was to determine if any of the documented requirements could be reduced to the point where commercially available equipment could be utilized.

There has been a traditional conflict within the Army telecommunications community regarding the use of commercial equipment in the tactical telecommunications environment. The primary reason has been the fact that the functional performance characteristics of commercial equipment have not met the specified tactical functional requirements. Additional factors have been the rigorous physical environment of the battlefield (e.g., shock, vibration, temperature extremes), the anticipated effects of the nuclear battlefield (e.g., EMP, radiation), and logistics support at distances remote from the factory location. None of these factors must be taken into account for equipment installed in the commercial environment.

The advantages of acquiring available, commercial telecommunications equipment are obvious. First, of course, is the fact that the equipment is available. Second, commercial equipment is built with the latest technology, unlike military equipment, which is built typically with technology at least one or two generations older. Relatively lower equipment costs are also a factor. This results from the large number of commercial items produced for the marketplace as well as the lesser set of performance characteristics to be satisfied. Reducing the specified performance characteristics of tactical equipment would allow commercial equipment to compete for application and, in turn, enable the military to benefit from the advantages that commercial equipment may offer. The Army has always procured some telecommunications equipment directly from the commercial marketplace. Recent initiatives are intended to increase the quantity and types of that equipment when it has been determined to be appropriate.

In addition to the needs of the tactical telecommunications community, there are significant activities that must be pursued in the nontactical military environment. The successful pursuit of these activities is essential in order to provide requisite support to the doctrines to be applied to the battlefield. Base communications (post, camp, station) are undergoing an upgrade to provide modern voice communication capabilities as well as those facilities necessary to handle the large amount of data traffic that is presently being generated and that is expected to increase significantly in the future. Computer networking, electronic mail, and other capabilities are being added to the traditional voice and message networks previously available. New private branch exchanges (PBXs) are being installed for voice and data service along with both the procurement and leasing of digital transmission facilities. These activities are occurring both in the United States and overseas. The highest authorities are firmly convinced that adequate, modern communications facilities are essential to the maintenance and success of a modern army. Since it is the case that Army personnel spend most of their time in garrison, the communications capabilities provided to the forces while in garrison must support their operations.

Communications equipment and systems for the strategic military environment will be and are being developed to support the new tactical doctrines as well as to be responsive to the defensive and offensive policies of the highest national authorities. Programs for the modification or replacement of communications facilities for the Joint Chiefs of Staff (JCS), National Command Authorities (NCA), and the Theater Commanders-in-Chief (CINC) are going forward. These programs include facilities for transmission, switching, terminals, and security.

In each of the three major telecommunications communities (tactical, base, strategic), care must be taken to ensure compatibility and interoperability of the equipment and systems at the interfaces among the communities. These aspects are addressed by careful definition and implementation of the architecture of each telecommunications system. The designers are working together to ensure that communications from ''the White House to the foxhole'' can be established and maintained.

2.4   THE IMPACT OF COMMUNICATIONS TECHNOLOGY ADVANCES

In the area of transmission, much effort is being initiated in the areas of high frequency (HF), very high frequency (VHF), and extremely high frequency (EHF). HF has been deemphasized in the past in the military but there is a recognized need to add modernized HF equipment and systems to the field. VHF technology is moving ahead rapidly and, as with HF, the military is taking advantage of the technological advances becoming available. EHF is a new and expanding field. It is finding applications in the terrestrial-based and space-based (satellite) environment. These technologies will be discussed in more detail in Chapter 4.

In the area of terminals, "user-friendly" is a term that is gaining popularity in all user communities. Voice terminals, facsimile terminals, optical-character readers, message terminals, and computer-imbedded terminals are being developed or acquired that provide users with increased capabilities without requiring extensive training for their operation. Size and weight of the items are being reduced as a result of the application of the newest technologies.

In the area of switching, the word "distributed" is finding favor with architects and planners. Large and heavy circuit and message switches will be used until they no longer can be maintained and supported, but it is unlikely that they will be replaced with more modern replicas. Instead, different switching techniques, such as packet switching, will be implemented. One may hypothesize that switching as we know it today will disappear from the tactical telecommunications environment. The functions will continue to be performed but they will be absorbed into other equipment assemblages such that, in the future, we will not see hardware or equipment whose sole purpose is to provide circuit or message switching.

All of the new equipmemt and systems must possess certain characteristics and qualities. In the areas of reliability and maintainability, the ideal is "high" reliability and "throw-away" maintenance. This would provide for significant reductions in training efforts and the expense and burden of creating and maintaining vast amounts of maintenance and repair documentation. Such a situation appears to be feasible but it can be attained only if the replacement parts are relatively inexpensive.

The new equipment must provide for adequate security (encryption) for the information being processed and exchanged. The key word is "adequate," for volatility of the information should be considered when the degree of security protection is being determined. In the past, the monetary cost of providing security capabilities was a deterrent to their inclusion in all areas of need. Perhaps volatility, along with other criteria, can be used in determining who and what gets protected.

Protection against electronic warfare (EW) and electronic countermeasures (ECM) has been a severe burden for developers of telecommunications equipment and systems for a long time. Furthermore, the determination of the threat to be encountered in any particular environment has always been a "catch-up" process. It is likely to remain so for the future. However, the equipment planned to be developed will contain capabilities heretofore considered impractical to implement. This is an area that deserves more and better attention by the entire telecommunications community.

The tasks ahead are formidable and exhausting, but the benefits to be obtained are essential and rewarding. An M1 tank battalion deployed on a battlefield without adequate communications can be likened to a modern aircraft carrier flotilla with inoperative steering mechanisms. The solutions to a myriad of problems are being found and implemented at an affordable cost with technology that is available. Inertia is being overcome and the momentum is increasing.

It has been pointed out that the Army has some unique communications opportunities. The same is true of the other services as well. The Air Force has unique avionics situations in which rapid, secure data transmission must take place between aircraft or from aircraft to ground with a low susceptibility to jamming and high probability of data delivery. This must all be performed in a hostile battlefield situation.

The Navy must communicate within its task forces, to other task forces, and from sea to land under hostile conditions that are unique to its missions. Submarines must stay submerged and quiet. Aircraft carriers must provide for reliable and survivable air support and battleships and destroyers must provide fire support.

A common ingredient among all the services is that they are all evolving to a digital communication system that has the following basic ingredients, as depicted in Figure 2.2:

- Information source encoding
- Encryption
- Channel encoding
- Multiplexing
- Frequency spreading
- Multiple access

The shaded boxes indicate that not all systems perform all of the functions. However, for contemporary communications systems utilized by the services, most of the subsystems are utilized.

Modern communications systems are depicted by sophisticated digital processing utilizing fast Fourier transforms, key distribution systems, new encoding techniques, efficient modulation techniques such as differential phase shift keying (DPSK), quadrature phase shift keying (QPSK), antijam techniques, such as spread spectrum (which includes direct sequence encoding, frequency hopping, and time hopping), multiplexing involving voice/data integration and multiple access techniques involving time division multiple access (TDMA) and burst time division multiple access (Burst TDMA).

2.5   COMMUNICATIONS FOR THE LAND BATTLEFIELD (A SPECIFIC EXAMPLE)

Before service communications systems are discussed in detail, it would be useful to consider a specific example of how the critical C^3I requirements can be satisfied in a battlefield situation.

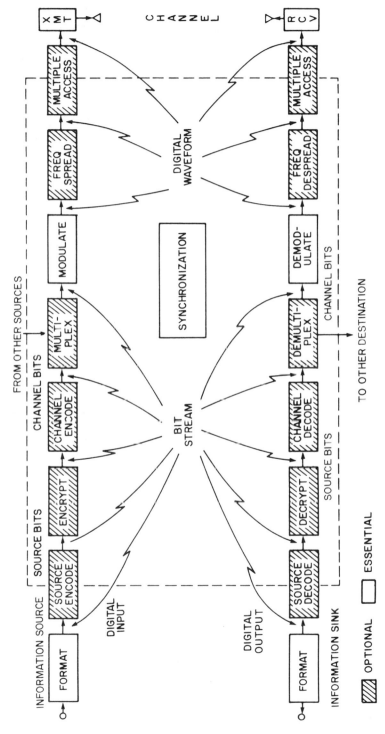

Figure 2.2 Digital Communication System.

Although the Navy and Air Force are part of any task force, as in the Grenada conflict, this example addresses the Army and some unique communications requirements.

A typical Army communication configuration would consist of an echelon above corps (EAC), corps, division, brigade, and battalion level C^3I structure, as shown in Figure 2.3 [7,8]. The most mobile equipment would be down at the brigade and battalion level in support of command centers. At the division level there would be less mobile equipment such as unit level switches, troposcatter radios, small satellite terminals, and mobile subscriber radios. At the corps level would be transportable large switches, satellite terminals, and line-of-sight (LOS) and troposcatter radios covering a larger area and providing connectivity to the division and EAC areas. The EAC area includes fixed communications capability such as switches of the Post Telephone and Telegraph (PT&T) in Europe. Figures 2.4 and 2.5 show the command center nodal relationships at the various echelons and the intelligence sensor and processing systems, respectively, to complete the C^3I picture. More details on these functional relationships can be found in the latest Army, Air Force, and Marine Corps field manuals and doctrinal positions.

Since the heart of the tactical war will be at the division level, that is the level that will be focused on in this chapter. Figures 2.6 and 2.7 show the physical equipment and geographic distribution for typical tactical communications setups.

Figures 2.8, 2.9, and 2.10 show typical models of communications equipment planned for the late 1980s and how they may be interconnected. The theme throughout the division tactical area will be mobility, jam resistance, and responsive data distribution. As is shown, the battlefield will be comprised of the TRI-TAC family of equipment, which is discussed in more detail in Chapter 3.

The division will most likely utilize mobile equipment including radios, switches, and terminals. A complete description of all of this equipment is beyond the scope of this book. However, by way of illustration, a simplified centralized communication network using mostly off-the-shelf commercial equipment is discussed for army tactical communication at the division level. Only one cell consisting of four typical nodes, e.g., a Tactical Command Post (TACCP), an All Sources Analysis Center (ASAC), a Divisional Command (DISCOM), and a Tactical Operations Center (TOC), is considered in detail (Figure 2.8). The whole network may consist of a number of similar interconnected cells.

The typical tactical network is shown in Figure 2.11. It consists of four nodes communicating with one another through a relay node on a TDMA basis. Assumed distances between the various nodes are also shown in Figure 2.11.

It is assumed that each of the nodes is producing information at the rate of 200 Kbps. This information is obtained under the assumption that each node is serving five subnodes, each consisting of:

- One 16 Kbps voice circuit
- One 2.4 Kbps data source
- One 16 Kbps facsimile device

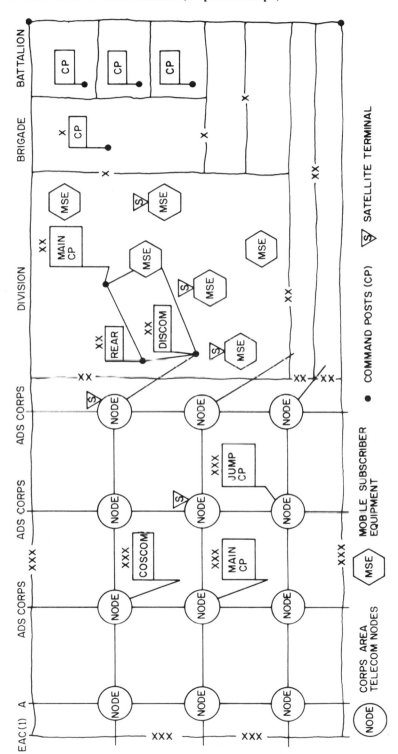

Figure 2.3 Telecommunications on the battlefield.

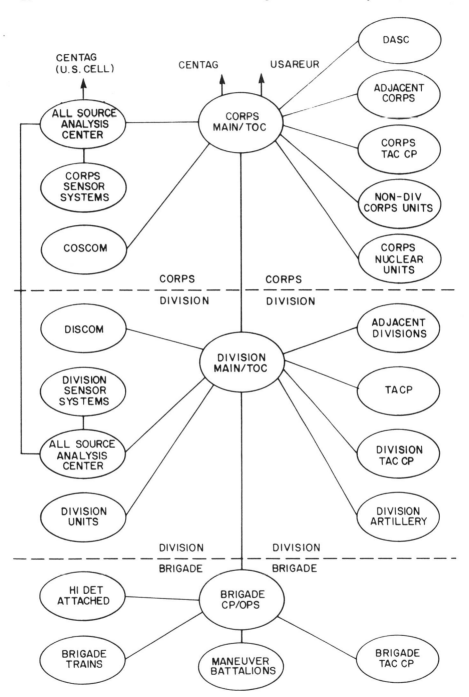

Figure 2.4 ATACCS nodal relationships for conventional war analysis.

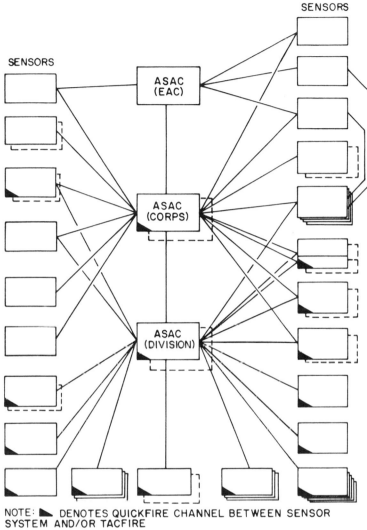

NOTE: ◣ DENOTES QUICKFIRE CHANNEL BETWEEN SENSOR
SYSTEM AND/OR TACFIRE

ASAC - ALL SOURCE ANALYSIS CENTER

Figure 2.5   ASAC-sensor system connectivity architecture.

Also, each node is allowed 28 Kbps for position and status report.
Each of the four nodes consists of:

- A number of small microcomputers
- Associated memory and other peripherals
- A TDMA terminal with modem

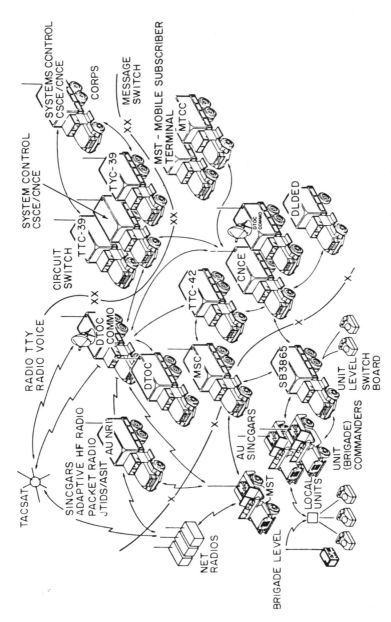

Figure 2.6 Typical division main node telecommunications equipment.

- An RF interface circuit
- A portable antenna

A typical configuration is shown in Figure 2.12.

The processor subsystem consists of a number of small microcomputers (comparable to the Apple II series) with associated memory and other peripherals. The relay node contains, in addition to the equipment at the other nodes, a TMDA burst subsystem as shown in Figure 2.13. The TDMA burst subsystem performs the functions of:

- Network control—protocols relative to setting up, maintenance, and breaking up of the network
- Time division multiplexing
- Time division demultiplexing
- Error control

The network facilitates communication among the four nodes via the relay node, which is to be placed in an elevated position so that the other nodes can communicate with it in the line-of-sight (LOS) mode. The TDMA burst subsystem works on a TDMA frame basis. A control frame consists of a number of data frames (the number of data frames in a control frame is equal to the number of nodes in the network). A master frame is formed out of a number of control frames. The burst subsystem receives the requests for service from the nodes and assigns slots for transmission to and from the nodes. Each node transmits/receives its information on the assigned slots. The slots are assigned through the Network Operation Center (NOC). Burst and requests for slots are transmitted in the ranging/control burst in the overhead slot of the data frame.

2.5.1   Network Performance

In this section, a number of questions relative to TDMA systems performance are addressed; they include the data rate, data/voice mixture allowable by this network, the frequencies to be used, jamming considerations and advantages.

2.5.1.1   Attainable Data Rate

In the present form, the maximum data rate that any terminal can transmit is 250 Kbps. It is seen from our previous discussion that this data rate is sufficient for the proposed system.

2.5.1.2   Voice/Data Mixture

The TDMA system can handle any mixture of voice and data. However, for this proposed network, each node is to have a mixture of approximately:

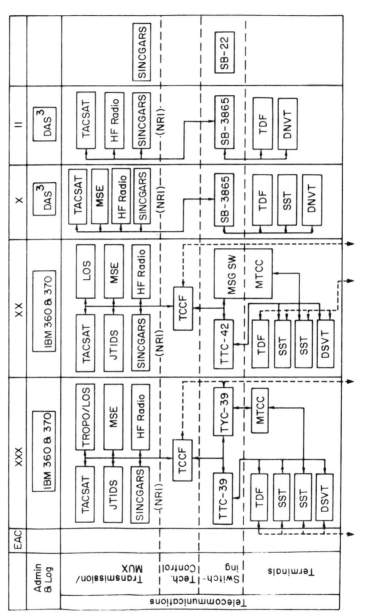

Figure 2.7 Army communications.

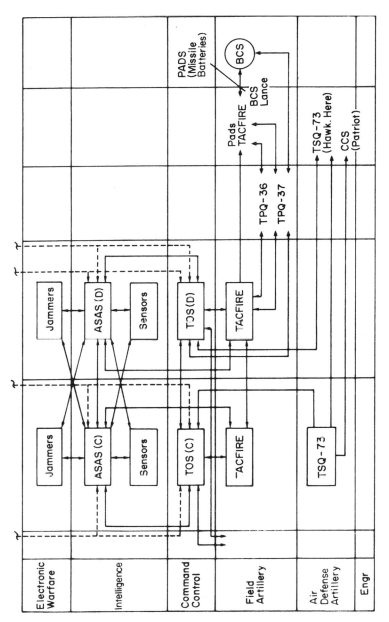

Figure 2.7 Army communications (continued).

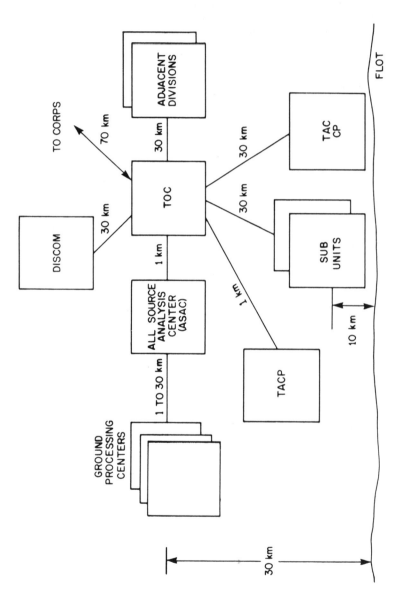

Figure 2.8   Division model.

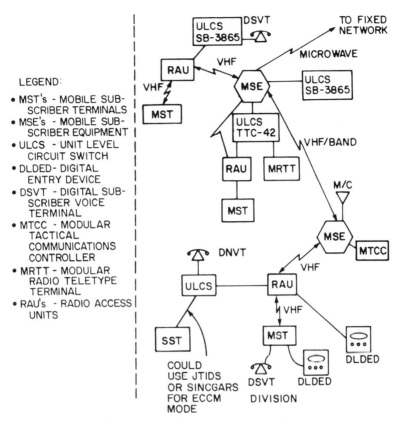

Figure 2.9 Division network model.

- 40 percent voice (five 16 Kbps voice circuits)
- 60 percent data (five 2.4 Kbps data devices, five 16 Kbps facsimile devices, and 28 Kbps for position and status report)

2.5.1.3 Frequency Selection

Transmission frequencies in the 800–1000 MHz (UHF) range are recommended, e.g., a 75 MHz bandwidth in the 806–881 MHz range. Two separate carrier frequencies or one frequency within this band can be used for the forward and the return paths between the relay station and each of the nodes.

These frequencies are suitable because:

- They have been successfully used in other mobile communication systems and, hence, off-the-shelf equipment may be available.
- These frequencies are suitable for distances of 10–50 miles.

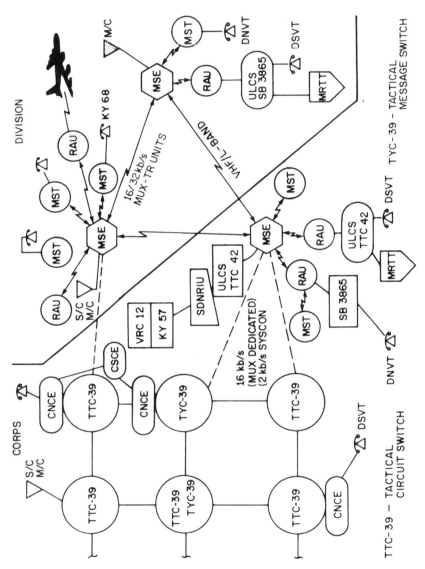

Figure 2.10 MSE network detail.

TTC-39 — TACTICAL CIRCUIT SWITCH

TYC-39 — TACTICAL MESSAGE SWITCH

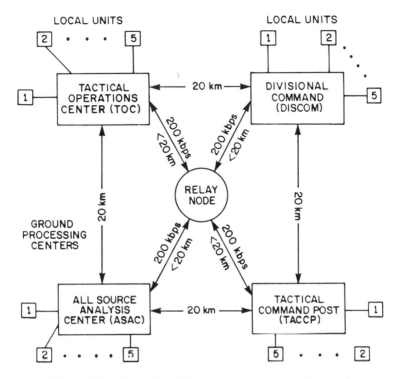

Figure 2.11 A model of the proposed experimental network.

- Transportable antennas can be used (parabolic reflector antennas with horn feed at the focus range from 3 to 6 feet in diameter for mobile equipment).
- Reasonable antenna directivity is obtained at these frequencies.
- Broad information bandwidth becomes available.
- Higher frequencies may produce higher propagation loss, fading, and receiver noise.
- At lower frequencies, more atmospheric and man-made noise occur.
- At frequencies above 10 GHz appreciable loss is introduced by rain, fog, and snow.

2.5.1.4 *Jamming Considerations*

Jamming and interference problems can be taken care of (if the BW is high enough) by using:

- Spread-spectrum and/or encryption Comsec equipment
- Frequency hopping, if necessary

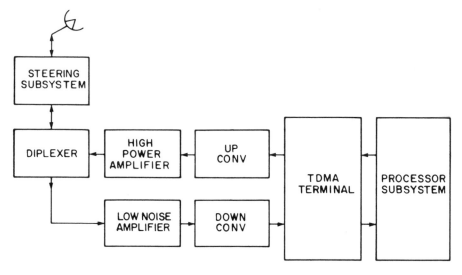

Figure 2.12   A typical node.

- Highly directional antennas (array antennas) (needed to reduce the amount of information interception by the enemy)
- Automatic repeat request error control protocol

2.5.1.5   *Advantages of TDMA Systems*

Typical advantages of TDMA systems over FDMA systems are:

- Terminal equipment is cheap and compact.
- Amplitude linearity of the links is unimportant.
- Individual users can be readily dropped and inserted.
- Demand assignment TDMA allows better utilization of the channel bandwidth.

More specifically, the TDMA proposed for this network example has the following advantages:

- Up to 240 terminals per net
- Up to 110 users per terminal
- Full demand assignment
- Fully redundant modem and controller
- Arbitrary voice/data mixture
- Forward error correction (optional)
- Wide range of user I/O ports

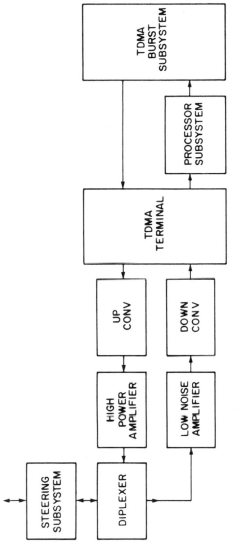

Figure 2.13 The relay mode.

- Low overhead due to improved method of synchronization
- Total digital demodulator
- High efficiency of TDMA frame (typically 99 percent)
- Frequency hopping (when implemented) providing jam resistance and high throughput
- Less cost than other comparable systems

In this initial system, frequency hopping is not included. However, it can be implemented, if necessary, to combat jamming.

With the advent of low-cost, reliable radio frequency and microprocessor electronic devices, it appears that off-the-shelf equipment is ready to be utilized on the battlefield.

The example just discussed indicates a typical state-of-the-art communications solution for the Army at the division level. In this approach, commercially available microcomputers are utilized in conjunction with TDMA LOS radios to provide for connectivity to command centers in a stressful situation witnessing physical damage and jamming threat. All of the services have a need for unique communications solutions to meet their requirements in the immediate future. The remainder of this book is devoted to practical techniques for providing improved communications capability and thus the much-sought "force multiplier."

2.6 REFERENCES

1. Edwards, E.C., "Strategic/Tactical Interface Connectivity." *IEEE Communications Magazine* 21, 4 (July 1983).
2. Buick, D.B. and Ellersick, F.W. "Future Air Force Tactical Communications." *IEEE Transactions on Communications* (September 1980).
3. La Vean, G.E. "Interoperability in Defense Communications." *IEEE Transactions on Communications* (September 1980).
4. Wentzel, Hingorani G.D. "NATO Communications in Transition." *IEEE Transactions on Communications* (September 1980).
5. Mannel, W.M. "Future Communications Concepts in Support of U.S. Army Command and Control." *IEEE Transactions on Communications* (September 1980).
6. Blackman, J.A. "Switched Communications for the Department of Defense." *IEEE Transactions on Communications* (July 1979).
7. Ricci, F. et al. "Army Command and Control Master Plan (AC²MP)." *IBM* (1979, (10 vols.).
8. Ricci, F.J. *IEEE Communications Magazine* (July 1983). Special Issue on Military Communications.

Chapter 3

U.S./NATO SYSTEMS

3.1 FIXED PLANT (EUROPEAN THEATER)

In the European theater there are a number of distinct, independently developed communications systems that must be interfaced and interoperated to support joint cooperative military action between the United States and its allies and to support connectivity between the intratheater tactical commanders and the high strategic level command.

As the NATO and U.S. forces communications network systems in Europe evolve, there will be a greater need for subscribers in one network to communicate with subscribers in another network. Figure 3.1 shows a typical configuration of the interfacing that will be required in Europe. The NATO network will utilize resources of the NATO Integrated Communications Systems (NICS).[1] The Defense Communications Systems (DCS) will utilize resources of the present and future AUTOVON/AUTODIN and European Telecommunications Systems (ETS) in addition to special networks and resources of the National Post Telephone and Telegraph companies (PT&Ts). The tactical networks will make use of the proposed Integrated Tactical Army Communication System (INTACS)[2] systems, enhanced by elements of the TRI-TAC family of secure voice, data, and facsimile subscriber equipment switched through the PBX's unit level switches and the TRI-TAC AN/TTC-39 and AN/TYC-39 backbone switches. The interconnection of these various networks requires that they be interoperable. That is, a subscriber in one network must be able to communicate with a subscriber in any other network [1].

---

[1]Including NATO-owned transmission systems, such as the Allied Command Europe (ACE) high troposcatter system, a satellite communications system, and various microwave radio relay and cable systems (e.g., European Tactical Communications).

[2]Integrated Tactical Communication System (INTACS) presently being implemented by the Army, Air Force, and Marines, discussed in more detail in Chapter 4.

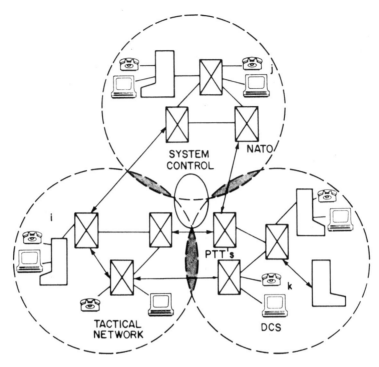

Figure 3.1   Subscriber-to-subscriber interface through theater network.

3.2   A TYPICAL TELECOMMUNICATION PROBLEM

This section will address a typical problem of a voice subscriber, i, in a tactical network desiring to call a voice subscriber, j, in a NATO network over the operational tactical and strategic voice networks [2,3].

As we have indicated, to accomplish interconnectivity, the European theater tactical and strategic networks must be interoperable; that is, the numbering plans, signaling, security techniques, and system control must be compatible. Interoperability encompasses many technical factors. For a number of years, interoperability groups have been meeting to ensure that all forms of communications can take place between the diverse systems in Europe. For example, the tactical community has cooperated with NATO countries in writing standards, called STANAGS, to ensure commonality between NATO switches such as the Initial Voice Switched Network (IVSN) and the TRI-TAC family of equipment. The TRI-TAC family of switches has been designed to interoperate with both TRI-TAC switches and NATO switches. That is, a TTC-39 circuit switch can implement the proper signaling and numbering plan to provide connectivity to subscribers.

As shown in Figure 3.2, the techniques of signaling and numbering utilized by subscriber i must be similar to those utilized by subscribers j and k. The dashed line in Figure 3.2 shows the signaling from subscriber i to subscribers j and k on a conference call. The technique for signaling and numbering must be standardized to the extent that, for example, dialing YYY-NNX-XXXX in one system can be recognized by other systems.

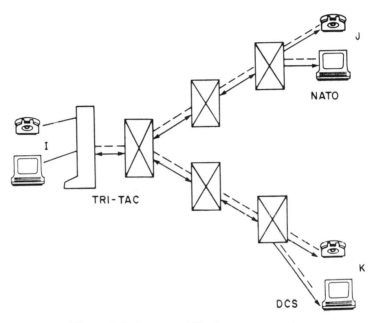

Figure 3.2   Interoperability between networks.

In order to achieve a level of interoperability that permits reconstitution of any valuable communications assets, the routing and system control techniques utilized must be compatible. Each switch in the network must be able to recognize the calling and called subscriber and provide the appropriate routing techniques to connect the subscribers with each other.

The following sections will trace a connection of subscribers, i, in a tactical network to a subscriber, j, in a NATO network by considering:

- Subscriber features
- Network numbering plans
- Network switching and routing
- Network interfaces
- Network services

3.2.1 Subscriber Features

The voice subscribers on each of the inherently incompatible networks, NATO
Integrated Communications System (NICS) and INTACS (the TRI-TAC family)
of switching equipment, will be able to communicate with each other because
of the capabilities built into the AN/TTC-39 circuit switch subsystem and/or the
AN/TTC-38 analog automatic switch subsystem, unless the interconnection is
prohibited because of unique and incompatible security devices or call service
classmarks. The subscriber-to-subscriber interconnection would be made through
the switches, over an analog interswitch trunk, capable of providing a four-wire
line termination to the NATO access switch subscriber (part of the NTCS Initial
Voice Switched Network (IVSN)) and either a two-wire or four-wire line termination to the tactical network voice subscriber. If necessary, the NATO switch
could provide the interconnection through a EUROCOM Interface Facility (EIF),
which is also compatible with the AN/TTC-39 switch.

The interfaces to the NATO communications system from the AN/TTC-39
switch are accomplished through a standard interface. These interfaces cover the
electrical parameters, supervision signals, and address signals common to both
the AN/TTC-39 and NATO common standards, and vice versa. The AN/TTC-
39 will perform the interfacing using a NATO interface unit capable of terminating eight voice channels. There can be two interface units per switch (sixteen
channels), which can be functionally grouped into up to four interface trunk
groups. This will allow the AN/TTC-39 switch to interconnect to up to four
NATO nodal switches (for alternate routing, survivability, etc.) (Figure 3.2).
The functions of the interface unit will be to:

1. Convert SF supervision and signaling (from the AN/TTC-39) to DC (four-
   wire) supervision and signaling on the NATO side, and reverse.
2. Provide local or remote interfaces interconnected via cable or other transmission medium.

Figure 3.3 shows the interfacing arrangement between INTACS and NICS
IVSN. More details are shown in Figures 3.4 and 3.5. The NICS IVSN will
interface with tactical switched systems at its Standard Interface Equipment
(SIE). The SIE provides for a nominal four-wire 4 KHz analog subscriber interface. This equipment will normally be located at a NATO access switch or
at a PABX.

3.2.2 Network Numbering Plans

The INTACS AN/TTC-39 circuit switch can provide three types of numbering
plans:

1. North American ten-digit AUTOVON/BELL System compatible
2. Tactical—as specified by MIL-STD-188C— a seven-digit PR-SL-XXX plan
   (PR = area code, SL = switch code, and XXX = subscriber extension)

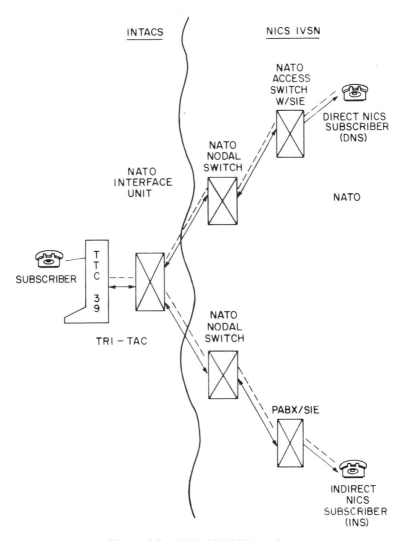

Figure 3.3   INTACS/NICS interface.

3. NATO EUROCOM System Access—up to 16 subscriber-dialed digits

The basic AN/TTC-39 numbering plan, exclusive of prefixes, is of the form NYX NN XXXXX, where NYX represents a set of national U.S. tactical and strategic area codes and subscriber numbers. The NATO (and EUROCOM) access code is of the form 9YX XXX XXXXXXX. The 9YX prefix is used for special handling such as precedence traffic mode selection of conferencing. The first three-digit prefix in the address code (XXX) is used for non-NICS networks, or area codes, and if necessary for future expansion of the NICS. The seven remaining digits (XXXXXXX) form the subscriber directory number, which is

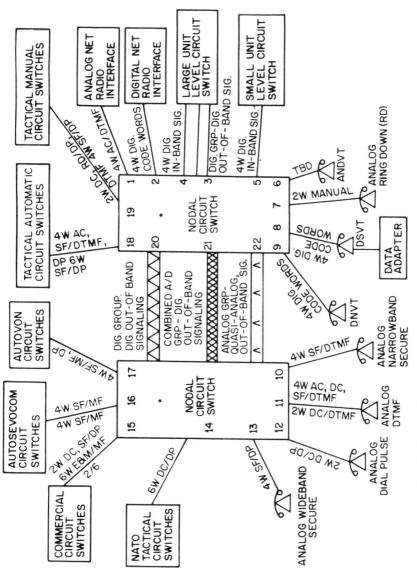

Figure 3.4 Nodal circuit switch interconnectivity capabilities.

*Numerals identify interface types indicated by corresponding interface reference numbers.

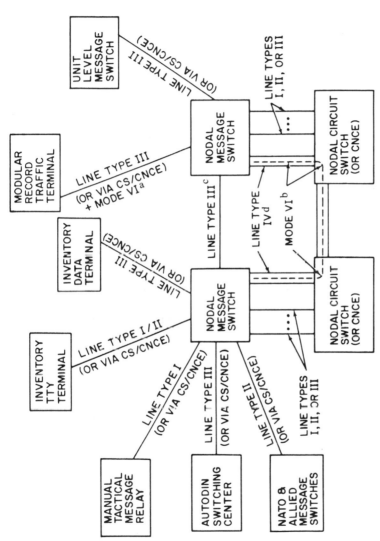

Figure 3.5 Nodal message switch interconnectivity capabilities.

[a] The Mode VI MRTT access must be via a nodal circuit switch or CNCE for inclusion in a Line Type II DTG.

[b] The Mode VI interconnection between MSs must be via nodal circuit switches or the CNCE.

[c] A Line Type III Interface is available on both a DTG and non-DTG basis.

[d] This is any DTG between nodal circuit switches or CNCEs and need not contain only or all of Line Type IV DTG.

designed to accommodate all existing three-, four-, and five-digit PABX extension numbers as the last part of the seven-digit NICS directory number.

3.2.3 Network Switching and Routing

The switching and routing of a call from a TRI-TAC circuit switch to a corresponding NATO switch, including transit calls to NATO/EUROCOM subscribers and to U.S. subscribers served by NATO/EUROCOM systems, is valid for the following scenarios [4]:

1. The called NATO/EUROCOM subscriber is located within the NATO/EUROCOM network.
2. The called NATO/EUROCOM subscriber is under command of a U.S. tactical force.
3. The called U.S. subscriber is under command of a non-U.S. NATO/EUROCOM tactical force.
4. The calling and called subscribers are located in NATO/EUROCOM systems and the U.S. system provides a connection between the originating and terminating system for call transit. Similarly, the NATO/EUROCOM system may provide a transit connection for calls between two U.S. systems.

To reach the called NATO/EUROCOM subscriber, AN/TTC-39 subscribers, including long loop subscribers, will dial the specific international access code (9YX)—also called the International Access Prefix (IAP)—followed by a seven- to ten-digit subscriber number plus the end-of-dial (EOD) where required. The AN/TTC-39 will then route the call on the basis of the International Access Prefix (IAP) plus the National Area Code (NAC) if included in the detailed address and listed in the alternate area routing table. Calls for an unlisted national area code will be routed on the basis of the international access prefix. The NATO systems will be classmarked for either three- or six-digit routing capability.

To reach the called NATO/EUROCOM subscriber under U.S. command, the U.S. subscriber will dial the international access prefix followed by the assigned U.S. area code of XXX for the designated subarea code, the seven-digit subscriber number, and the EOD. Call processing and routing will be equivalent to that for access to foreign networks.

To reach a U.S. subscriber of a U.S. force under NATO/EUROCOM command and accessible through the NATO/EUROCOM network, the U.S. subscriber should dial the U.S. area code and seven-digit subscriber number of the U.S. national access prefix (e.g., 914). AN/TTC-39 call processing and routing will then proceed as specified for calls to NATO/EUROCOM subscribers.

The U.S. subscriber may also dial a prefix to the dialed NATO/EUROCOM address to specify special service requests, as for intra-TRI-TAC system calls.

3.2.3.1 Requirements and Constraints

The AN/TTC-39 will be capable of accommodating all 20 9XY international access prefixes and up to 30 combined 9YX XXX codes. It will be capable of inhibiting NATO/EUROCOM transit calls, i.e., calls from one NATO national network to another non-U.S. NATO national network, by using a zone restriction classmark. In addition, other service restriction classmarks generally associated with subscribers' or PBX access trunks will be able to be applied to the NATO/ EUROCOM interface trunk groups.

In AN/TTC-39 precedence handling, a preemptive search will be initiated if a call is of a precedence higher than ROUTINE and an acceptable idle trunk that does not increase the satellite path delay beyond the established threshold parameters is not found in the initial search of the primary and alternate trunk groups. The preemptive search will scan, in order, the primary and alternate trunk groups for the desired type of channel (analog or digital). The lowest precedence level trunk of all trunks in the primary and all alternate trunk groups acceptable and encountered first in the scan will be used to forward the call. To be acceptable it must be of lower precedence than the preempting call request, must satisfy the path delay, transmission restrictions, and security criteria, and must not have been previously marked as unavailable. If no preemptable acceptable trunk is found, the ALL TRUNKS BUSY (ATB) shall be returned from a noncontrolling switch and/or from the originating switch of the subscriber or to the equivalent inventory switch interface trunk group as appropriate. A controlling intermediate switch will return BUSY RELEASE if an acceptable preemptable trunk is not found. Any intermediate switch receiving BUSY RELEASE will pass it on toward the originating switch and release the seized or reserved trunk.

AN/TTC-39 subscribers may use the X EOD prefix, consistent with their terminal capabilities, to request special services. The AN/TTC-39 may receive the route code R on AUTOVON interface trunks. In addition, calls from NATO and calls from EUROCOM may include special service prefix digits. NATO call prefix digits are in accordance with NATO common interface standards. EUROCOM interface prefix digits are shown in Table 3.1. Special prefix digits will be included in the interswitch trunk CALL INITIATE signaling message. The interswitch prefix and address format is shown in Table 3.2

The AN/TTC-39 will be capable of operating in an all-digital as well as mixed analog/digital network, which includes AN/TTC-39s, AN/TTC-42s, SB-3865s, and inventory switches. It will be capable of performing routing and associated control functions as specified in U.S./NATO standards. The AN/TTC-39 routine actions for a particular call will be consistent with the role assumed for the given call, i.e., originating switch, intermediate or medial switch, parent switch, proximal and digital area interface switch role, etc., and will satisfy the following basic requirements and constraints:

- The AN/TTC-39 switches may be interconnected by all digital, mixed analog and digital, and in some cases all analog trunk groups. Mixed analog/digital

Table 3.1 Prefix code assignment

| | 16-Digit Prefix Plus Address Field | | | |
| | 3-Digit Prefix | | 13-Digit Address | |
Originating System	Digit 16	Digit 15	Digit 14	13 to 1 /6
AUTOVON	Blank	Digit 1*	Digit 0-9 /5	0-9
NATO	Blank	Digit 2*	Digit 0-9 /4	
TRI-TAC	Blank	Digit 3**	Digit 0-9	0-9
EUROCOM	Not Blank***	Any 4 Bit Code***	Any 4 Bit Code***	0-9

*To be inserted by the interfacing proximal TRI-TAC switch.
**To be inserted by the originating TRI-TAC switch (dial prefix).
***Received from EUROCOM.
/4 Mode prefix received from NATO.
/5 R-Code received from AUTOVON/SEVOCOM.
/6 Blank code is substituted for unused digits.

Table 3.2 Interswitch prefix and address format

| Prefix + area codes | | | | | | 7-Digit Subscriber Number | | | | | | | | | |
16th	15th	14th	13th	12th	11th	10th	9th	8th	7th	6th	5th	4th	3rd	2nd	1st
Additional Codes			B	B	B	N	Y	X	N	N	X	X	X	X	X
						or			P	R	S	L	X	X	X
						(B	B	B)	9	9	X	X	X	X	X
			9	Y	X	B	B	B	X	X	X	X	X	X	X
			N	Y	X	N	N	X	X	X	X	X	X	X	X
			X	X	X	X	X	X	X	X	X	X	X	X	X

International*   National*   —2 digits ——————— 5 digits—
Access   Access/area   —3 digits ——————— 4 digits—
Prefix-IAP   code or   —4 digits ——————— 3 digits—
                   international
                   sub-area code
                                      2-digit fixed   3, 4, or 5-digit
                                      directory   subscriber
                                      prefix or   address
                                      3- or 4-digit
                                      switch code

10, 9, 8 or 7-digit international address
(+B(s))

*Fields used for routing decisions.
B = blank character

trunk groups will be composed of trunks with the following combinations or transmission characteristics and security capabilities or restrictions: digital/secure (DS), digital/nonsecure (DN), analog/secure (AS), and analog/nonsecure (AN).

- When selecting a trunk for a "digital-preferred" call, the AN/TTC-39 will first search for an available idle digital channel in the primary trunk group. If necessary, the search will be extended over all listed alternate routes. Similarly, when selecting a trunk for an "analog-preferred" call received on an analog loop or trunk, the switch will first search all analog channels in the primary trunk group and the ordered set of alternate trunk groups for an idle channel. If none is found, the switch will convert the call to a digital call and will search the primary trunk group and all alternate routes in the order listed for an idle digital channel.

- In the process of selecting a trunk for a call marked "digital only" or "analog only," the search for an available channel will be limited to the type of channel specified. For selecting a trunk for a call with a nonroutine precedence level, a preemptive search will be performed. The type of trunk to be selected for preemption will be consistent with the type of trunk "preferred" or "required." In the case of a "preferred" type, preemption of a trunk of the "other" type will be exercised only if no preemptable trunk of the "preferred" type is available, and only if the switch has routing control as an originating or acting originating switch or a spill forward/medial switch.

- For FLASH and FLASH OVERRIDE calls, brute force preemption will be employed if no idle acceptable path is found in the primary trunk group; i.e., if no idle but an acceptable preemptable path is available in the primary trunk group, it will be employed prior to attempting alternate routing. Similarly, the second alternate trunk group will be searched only if there is no idle or preemptable, acceptable channel in the first alternate trunk group, and so forth. Within a given trunk group an idle acceptable channel will be seized if available prior to exercising preemption. If preemption is required, the lowest precedence level circuit available and acceptable shall be preempted.

- Analog data calls, i.e., externally encrypted data calls initiated by local analog subscribers using the "analog-only" dial prefix, will be routed over analog trunks only. Data calls from an interfacing AUTOVON switch (sending the R code "data") will be routed over analog channels where provided, that is, they will be routed as analog transmission "preferred" calls.

- The AN/TTC-39 will route calls originated from narrowband secure terminals (NBST) over digital trunks; that is, they will be routed as digital transmission "preferred" calls. The AN/TTC-39 will reserve an (external) NBST for the call if any of the following conditions is given:
  1. The originating local NBST loop is classmarked "NBST-encryption, analog path only."
  2. The local NBST subscriber dials the "analog-only" (3 EOD) prefix (the loop is classmarked "NBST, security preferred").

3. The call request is received on an AUTOSEVOCOM I interface trunk classmarked "NBST -encryption, analog path only."

3.2.3.2   Alternate Area Routing

When multiple paths exist to a distant area (area code or 9YX system), the routing selected will be based on the switch code (NNX) in the AUTOVON numbering plan (9NYXNNXXXXX); the subarea (PR or NN) or switch code [PRSL or NNX (X)] in the U.S. tactical number plan (NYX PRSLXXX or NYX NNXXXXX), and the national access code (NYX or XXX) in the international tactical number plan [9YX NYXXXXXXXX or 9YZ(XXX)XXXXXXX]. Eight alternate area routing tables, each assignable to any one area code (NYX) or international access prefix code (9YX), having the capability to store up to 10 switch codes [NNX(X) or PRSL], or subarea codes (NN or PR), or national access codes (NYX or XXX), will be provided. The alternate area routing table will identify a first preferred route and, if that has been previously assigned, a second preferred route capability. If the switch code, subarea code, or national access code does not appear in this table, or the preferred routes are busy, the call will revert to routing on primary and (up to five) alternate routes of the normal area code (NYX) or international access prefix code (9YX) routing procedures as specified in TT-B1-1103-0054. The AN/TTC-39 will have the capability to add or delete entries in these tables or entire tables via the SSF or upon receipt of the ALTERNATE AREA ROUTING TCCF directive. A DI-RECTIVE IMPLEMENTATION message or an ALTERNATE AREA ROUT-ING report will be generated and transmitted to the TCCF in response to a change having been made to these tables.

3.2.4   Network Interfaces

The external requirements encompass a set of equipment and procedural techniques that will be employed to facilitate intersystem connections and communications between subscribers of the tactical switched systems and subscribers of other systems, as indicated in Table 3.3. This subsystem will provide limited intersystem communications capabilities for voice, teletypewriter, facsimile, and/or data subscribers of the following external systems:

- Defense Communications System (DCS)
- Naval Telecommunications System (NTS)
- Designated U.S. Command and Control and data systems
- Allied tactical systems (near- and far-term)
- NATO Integrated Communications System (NICS)

The basic interface concept between tactical switched communications and external systems requires the use of distributed gateway interfaces between tac-

Table 3.3 Hybrid nodal circuit switch interface capabilities

Interface Type	Line Type	Bandwidth or Bit rate	Supervision	Signaling	Electrical Characteristics	Multiplex Formats
Analog loops	2W 4W 6W	Nominal 3kHz or 4kHz	AC, 20Hz RD, SF, DC	DTMF, Manual, DP	MIL-STD-188C MIL-STD-188-100	N/A
	4W	Nominal 50kHz	SF	DP	MIL-STD-188-100	N/A
Analog trunk	2W 4W 6W	Nominal 3kHz or 4kHz	AC, DC, SF E&M, RD	DP, MF 2/6, MF, DTMF, Digital-Out-of-Band	MIL-STD-188C MIL-STD-188-100	N/A
	4W	Nominal 50kHz	SF	MF, DP	MIL-STD-188-100	N/A
Digital loop	4W	16/32 kb/s	Digital Code Words IAW TT-43-9015-0046		Single Channel Digital Circuit	N/A
	4W	2.4 kb/s	TBD	TBD	TBD	N/A
Digital trunk	4W	16/32 kb/s	Digital in-band signaling IAW TT-A3-9012-0055		As for digital loop	N/A
Digital trunk group	4W	256-4608 kb/s	Digital Out-of-Band IAW TT-A3-9016-0056		Multi-channel digital group	ICD-003

tical nodal circuit and message switches and similar external switched systems. The tactical subscribers will be connected to external system subscribers via the distributed gateway switches. Interconnections between the interfacing switches will vary according to the level of compatibility between the switches, technical control, and Communications Security (COMSEC) and transmission equipment of the two systems. The following subsections describe the various generalized interconnectivity requirements between tactical switched and NATO systems.

The NATO Integrated Communication System (NICS) will contain various nominal 4 kHz analog transmission systems, an Initial Voice-Switched Network (IVSN), a pilot secure voice project (PSVP), and a Telegraph Automatic Relay Equipment (TARE) network.

3.2.4.1   Initial Voice Switched Network (IVSN)

The IVSN interfaces with the tactical switched system at the standard interface equipment (SIE), which provides for a nominal four-wire 4 kHz analog interface. Network-In-Dial (NID) and Network-Out-Dial (NOD) with DC supervision and DTMF (Discrete Time Multi-Frequency) in-band signaling will be used on this interface. The electrical interface between IVSN and the nodal circuit switch is to be in accordance with the B1 type interface.

3.2.4.2   Pilot Secure Voice Project (PSVP)

The Pilot Secure Voice Project (PSVP) is part of the NATO upgrade system intended to provide secure voice capability throughout the European Theater. The secure voice equipment used in the PSVP is called ELCROVOX.

The PSVP is an interim dedicated network consisting of ELCROVOX equipment for some 300 user locations. The equipment includes a pitch-excited channel vocoder capable of operating output rates of 1200, 1800, and 2400 bits/s. The PSVP is planned to be integrated into the IVSN.

3.2.4.3   Telegraph Automatic Relay Equipment (TARE)

The NICS will use the TARE network to provide store-and-forward message processing to dedicated subscribers. A nodal message switch of the tactical system will interface on a trunk basis with automatic relay equipment of the TARE network.

The interface trunk will operate at 600 baud and as an R community member, using ACP-127 NATO message formats. The tactical message switch will be capable of routing messages from U.S. subscribers to the designated TARE relay equipment and vice versa.

The transmission to TARE will be a nominal 4 kHz analog link. Transmission of unclassified (U), classified (Y), and routine unclassified (RU) (except designated RUXX) messages into the TARE network will be precluded. Only RU traffic to or from a TARE relay will be subjected to positive security screening to ensure that only NATO classification categories are allowed to cross the interface. Software safeguards will limit access to the NATO TARE interface to RU community subscribers only.

The interoperation of U.S. and Allied tactical circuit-switched systems will require both an analog interface for current inventory tactical equipment and pulse code modulation (PCM) systems, and a digital interface for future digital tactical equipment and digital group multiplexor (DGM) systems. The tactical switched system will provide a capability to interface with both types of systems.

The external system network typically interfaces via selected communications network control elements (CNCEs) and/or selected switches of the nodal switching systems. Connections to external systems can be achieved by loop, trunk, or dedicated circuit as the traffic requirements and the operational situations dictate. Many design parameters and characteristics of the external systems are compatible with the tactical system, thus permitting a high level of intersystem interoperability without the use of special interface facilities. Hence, the implementation of external interfaces will involve special interface facilities only in a limited number of applications. Normally, each implementation will be specified in directives issued by the Communications System Control Element (CSCE) staff via the CNCE.

A variety of functions will be performed by the external interface facilities in order to establish communications links into or through external systems. The set of functions associated with a particular interface facility will be tailored to satisfy the intended application. The following will typically be performed by the equipment of the system:

- Bit rate conversion
- Signaling conversion
- Address conversion
- COMSEC conversion
- Buffering for timing compatibility
- Precedence conversion

When a tactical network interfaces with external systems, a secure analog or digital voice orderwire and a secure TTY orderwire connecting the nodal control elements at each end of the interface are, as a minimum, required. Control of the transmission facilities will be provided by the system furnishing the facility.

3.2.5 Network Services

In the event that either subscriber needs additional services, whether in placing or during a call, both the AN/TTC-39 and the NICS IVSN offer the following compatible user services:

- Conferencing (progressive and preprogrammed)
- Automatic call transfer
- Abbreviated dialing
- Precedence calling
- Off-hook (direct access)
- Hot-Line
- Zone restriction (closed networks)

Each network offers, or will be offering, other special services that are available within its own network, and these could assist each subscriber with calling in the following areas:

- Security
- Camp-on-busy
- Idle line hunting
- Commercial network access
- Use of satellites for trunking

3.3 SUMMARY

The integration of U.S. and NATO telecommunications systems is an important ingredient for the defense of the European Theater. This chapter has provided a summary of key U.S. and NATO systems with a specific example of inter-connectivity and reference to:

- Subscriber features
- Network numbering plans
- Network switching and routing
- Network interfaces
- Network services

With some emphasis on tactical/strategic interfaces, a practical view of how U.S. and NATO systems can be integrated has been provided.

3.4 REFERENCES

1. Coviello, G.J.; Lebow, I; Pickholtz, R.L.; Schilling, D.L. *IEEE Transactions on Communications* (September 1980). Special Issue on Military Communications.

2. Ricci, F.J. "Mobile Telecommunications Solutions Using TDMA." RAMCOR Technical Report, July 1984.
3. Ricci, F.J. and Arozullah, M. "U.S./NATO Interface Systems." RAMCOR Technical Report, 1983.
4. Brand, R. "NATO Integrated Communications System Network Control." *IEEE Proceedings on Military Communications* (1983).

Chapter 4

ARMY, NAVY, AIR FORCE COMMUNICATIONS SYSTEMS

In response to unique ''battlefield'' telecommunications needs and in an effort to take advantage of advances in technology, many new systems are being developed for the Army, Air Force, Navy, and Marine Corps.

These new systems will evolve from the presently fielded systems to ones that will be responsive to the requirements of a mobile, fast-moving tactical force structure, responsive, that is, to both the conventional and nuclear war threats of physical damage and jamming. Although comprehensive architectures exist for the Air Force and Navy, this chapter will concentrate on the Army system architecture.

The Integrated Tactical Communications System (INTACS), an architecture proposed for implementation by the Army, consists of telecommunication equipment at the corps, division, brigade, and echelons above corps (EAC) levels. This system will include new VHF radios as well as elements of the TRI-TAC family of equipment and satellite systems. Some alternatives to the TRI-TAC system are also being investigated.

The other services, Navy, Air Force, and Marines will also utilize elements of the TRI-TAC family of equipment in addition to such systems as the Joint Tactical Information Distribution System (JTIDS), the Position Location Reporting System (PLRS), the Single-Channel Air-to-Ground Radio System (SINC-GARS), Mobile Subscriber Equipment (MSE), and Ground Mobile Forces (GMF) terminals to satellites. Those systems are intended to improve the security, jam resistance, connectivity, and capacity for voice, data, facsimile, and video communications.

The Army, Navy, and Air Force systems are discussed in the following sections.

4.1 ARMY SYSTEM ARCHITECTURE

The proposed INTACS system architecture is designed to serve as the basis for the Army's objective system of the future. The architecture utilizes the design specifications and operational characteristics of current equipment and equipment being developed under the improved TRI-TAC, MSE, and SATCOM programs to achieve the operational capabilities required to perform the Army's mission.

The architecture's design and flexibility is such that it can adapt to changing requirements and allow for the addition of new equipment developed in the future. The architecture provides the capability to connect today's tactical communications systems into the objective system while the development of an improved INTACS continues. The objective system involves a new organizational structure that provides the base design for a total telecommunications system.

4.2 ARCHITECTURE CRITERIA

Transition is the key word. The criteria for the Army transition is to develop a system that is fluid and dynamic in application. The system is constrained neither by time nor procurement strategy.

It will therefore be necessary and vital throughout the transition period to achieve and maintain the optimum system configuration for the higher priority users. Maximum use of second-level equipment will be retained by reissue of this equipment to lower priority users. Eventually, such reissued equipment will end its useful life with the lower priority users and will be replaced by new equipment issued from production lines at that time. Time phasing of equipment allocation will be critical. That phasing will be accomplished by means of automated computer programming that permits rapid changes and year-by-year adjustment.

4.3 CONCEPT OF OPERATION

The "Concept of Operation" to accomplish the transition to the INTACS Objective System is to:

- Restructure present units into the INTACS organizational structure for the objective system. To transition properly, the restructuring should be accomplished prior to the operation of the AN/TTC-39 and related equipment.
- Using the organizational structure of the objective system, build an integrated system utilizing current equipment (ATACS), new Army equipment (improved ATACS), and new objective system equipment (TRI-TAC, SINCGARS, PLRS/ JTIDS, MSE) when available, to provide a system that serves the present and future user with integrated equipment and interoperable systems.
- Establish a system test-bed to test the integrated system and its equipment.
- Establish teaching methods, training devices, equipment requirements, and teaching criteria as required, based on outcome of evaluations using the test-bed and computer simulation modeling.

- Teach the required knowledge to the commander, staff, supervisors, and equipment personnel so that they can function as a part of the new tactical communications world described in the INTACS Objective System.

4.4   TACTICAL SWITCHED COMMUNICATIONS

4.4.1   System Configuration

The Army System includes the full complement of switched and dedicated communication facilities required to accommodate traffic associated with ground-based elements of all services in a combat area. The system will provide an analog/digital, switched, multinodal communications network with end-to-end security for system subscribers. Transmission design is based on satellite and terrestrial transmission elements with flexibility to meet the needs of deployed forces. Store-and-forward message switches and circuit switches form the backbone of the switched system. The full range of switched communications terminal facilities (e.g., secure and nonsecure voice, message, facsimile, and data equipment) is provided for combat, combat support, and combat service support missions.

New systems will be established using tactical communication networks that consist of a hybrid mix of analog and digital switching, transmission, and subscriber terminal equipment. The ATACS equipment will consist of equipment that is now in the inventory, some of which will be used in its current form while some will be modified to ensure integrity of the network. This new phase will represent the initial step in the introduction of new digital equipment that will have an integrated COMSEC capability. The major elements introduced during this phase will include the circuit and message switches, secure and nonsecure digital telephones, a family of unit-level switches, digital multiplexing and transmission equipment, and nodal control equipment.

4.4.2   Nodal Configurations

The deployments will incorporate inventory equipment and the new INTACS equipment items. The inventory equipment includes inventory analog and PCM/TDM equipment. The incorporation of newly developed and inventory equipment into the system provides interconnectivity and interoperability possibilities. The operational and performance characteristics of the newly developed equipment are contained in their appropriate specifications.

The INTACS nodal system equipment can be classified into the general categories of Communications-Engineering (C-E) management system, transmission system, switching system and user equipment, plus the necessary interface components.

4.4.3   C-E Management Systems

The INTACS Objective System states the requirement for a set of facilities needed by system controllers for system planning, engineering, interconnecting, monitoring, testing, and managing designated portions of the tactical switched communications networks. When used collectively, the facilities will provide the capability for centralized system planning, engineering and control, and decentralized execution of system operation in order to support both component and joint service operations. These operations will range in scope from minor contingencies up to and including general war. Although not provided in the early stages, when the full complement of the Tactical Communications Control Facility (TCCF) hardware/software or equivalent is fully fielded, it will provide a computerized capability to facilitate communications system planning, system engineering, and system management on a near real-time basis. Presently, only the Air Force is procuring TCCF equipment.

4.4.4   TCCF Operational Communication-Control Concept

The communications control concept consists of four functional levels and a family of equipment that can be used by system controllers and equipment operators to exercise technical control of communications facilities and dynamic control of communications networks and to provide automated support for broad system planning and engineering. The control equipment will be capable of supporting component and/or joint systems and may be subdivided so that each subdivision is managed by designated elements operating at subordinate levels of control. The control equipment will provide for control of communications resources and automated management of network status information by rapidly acquiring, processing, and displaying data to appropriate level decision makers. The four levels of technical management and control are:

1. Communications System Planning Element (CSPE)
2. Communications System Control Element (CSCE)
3. Communications Nodal Control Element (CNCE)
4. Communications Equipment Support Element (CESE)

4.5   TRANSMISSION SYSTEMS (NODAL)

The technical control (TC) is the nodal focal point for all transmission functions of the switched communication system, providing the hub for all system applications. The TC is a computerized logic center that generates multiplex groups and channels for the transmission links and, as such, provides the nodal termination for all multiplex systems entering or exiting the node (Figure 4.2).

Nodal radio equipment consists of the AN/TRC-138 (Mod) for node-to-node links, the AN/TRC-174 and AN/TRC-173 for node-to-extension facilities, and

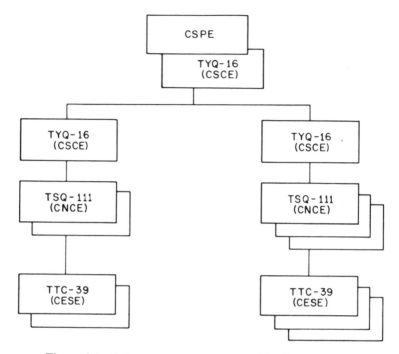

Figure 4.1   C-E management structure and hardware (TCCF).

the AN/TRC-175 (short-range wideband radio) for node to radio park links. Skip node (echelon) command and control and high priority users are provided for by TACSAT and tropo assemblages.

4.6   TRANSMISSION SYSTEMS (INTRANODAL NETWORK EXTENSION FACILITIES)

The intranodal or intrasite distribution of circuits and channel groups is accomplished through the use of network extension facilities. The facilities will serve individual groups of digital subscribers and provide a digital transmission capability by multiplexing channels into groups and combining groups. The network extension facilities will include a family of channel and group multiplexers and modems that facilitate transmission via single and multipair field wire, multiconductor cables, a multichannel coaxial cable transmission system, and multichannel line-of-sight radio transmission equipment. That equipment will provide the means for interconnecting nodal facilities, and internodal transmission equipment. All network extension facilities will be designed to carry both mission traffic and the associated control (orderwire) traffic.

The TC is designed to provide both nodal management and equipment applications for the system. The modular design of the TC allows a subsystem to

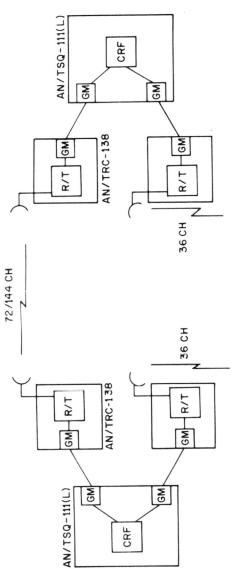

Figure 4.2 Transmission system (CNCE/DGM).

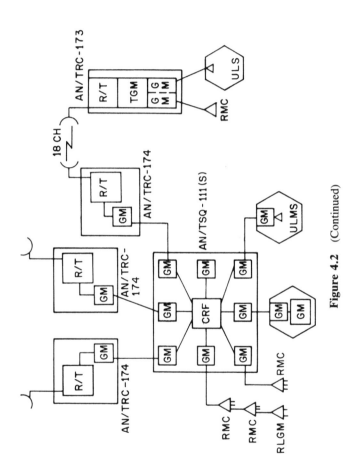

Figure 4.2 (Continued)

accomplish these tasks. The subsystems provide communication multiplexing and patching, fault detection and diagnosis, automation, COMSEC, and controller interface. The major subsystems and their functions are as follows:

- Analog Patch and Test—provides manual patching, line conditioning, automatic analog testing (AAT), manual (backup) testing, monitoring, and routing of all analog circuits
- Digital Patch and Test—provides line modems, manual patching, automatic patching (channel reassignment function (CRF)), and automatic digital testing (ADT) of all digital circuits
- Fault Detection and Isolation—provides processing of internal BITE data, telemetry fault and performance data, switch status report data, AAT test results, and channel reassignment status to detect and isolate faults in the nodal equipment
- Processor Subsystem—provides processing and storage of data; control of automatic testing, automatic patching, fault detection, and isolation functions; and processing for controller interface with the data base

4.7 DIVISION COMMUNICATIONS

A proposed division communications support system, discussed previously, is based on the INTACS concept and uses the INTACS system architecture to obtain and meet the requirements of the future in all four phases of division operation:

Phase I—Restructure of the division and deployment of improved ATACS equipment
Phase II—Integration of new equipment
Phase III—Deployment of division digital multichannel switching systems
Phase IV—Objective system application for division elements

The architecture utilizes the design specifications and operational characteristics of current equipment and the equipment being developed under the improved ATACS, TRI-TAC, SINCGARS, MSE, PLRS/JTIDS, and SATCOM programs to achieve the operational capabilities required to successfully perform the Army's mission.

The architecture's design and flexibility is such that it can adapt to the changing requirements and incorporate equipment developed in the future. The architecture presents criteria for connecting today's tactical communications systems into the objective system and maintaining the system. The system of the future is based on state-of-the-art equipment and a new organizational structure that provides the base design for a total telecommunications system that has total interoperability.

4.7.1 Phase I—Restructure Division

Only token restructure of the division can be accomplished during this period, but division communications equipment and capabilities will be improved to enhance division and separate brigade operations. Major improvements could consist of adding the TD-1065 to the existing PCM assemblages to provide a 16/32 kb/s digital capability, adding the TD-1069 to provide a 12-port data multiplexer for the divisional system, and adding TACSAT equipment for command and control. Restructure will be in accordance with FM 11-50 doctrine.

4.7.2 Phase II—Integration of New Equipment

To integrate the new INTACS equipment into the division an integrated system is created using both the existing equipment and the new equipment to accomplish the communications mission. Interoperability is achieved by integrating channels, groups, and A/D conversion. Integration is locked to the availability and amount of equipment received during each time frame. Major new equipment affecting integration and hybrid applications are:

- AN/TTC-42 Circuit Switch
- SB-3865
- AN/TCC-39A, AN/TYC-39
- AN/TRC-173/174 Radio Terminal and Repeaters
- GRC-103 Radio
- DGM Components
- COMSEC Equipment
- MSE
- SINCGARS

4.7.3 Phase III—Digital Multichannel Switching System

All major items of new equipment and user devices must be available to accomplish this task. ATACS equipment is deleted or modified for use in the new system. Analog subscribers left in the system are served through the AN/TTC-42, and CVSD capabilities. Major division elements are provided service into the corps system utilizing the mininode capabilities of the SB-3865 and AN/TCC-42, with the interconnection of the divisional switches and DSVT/DNVT subscriber and data elements.

In an effort to meet threat-driven requirements, reduce emitter concentration, and free the headquarters elements from the communications complex, the Phase III Alternative to the base doctrinal system is being developed. This alternative changes the divisional nodes into signal centers, adds three signal centers for balance, and area coverage allowing the headquarters and extension elements to satellite off the signal center freeing the users to move through the system with minimal communication recovery time.

A representative diagram of Phase III Alternative, Doctrinal System Modified, is shown in Figure 4.3 with the alternative system interconnection of nodes and extension elements.

4.7.4 Phase IV—The Objective System for Division Elements

Division elements are to get new FM radio systems (SINCGARS), new mobile radio-telephone systems (MSE), TACSAT terminals, high-speed message devices, and an automated data-distribution system (ADDS).

The division objective communications support system is based on the following concepts: mobile subscriber centrals operating in the VHF band to provide long-range coverage, wire subscribers served by unit-level switches and remote access units (AU), rapid secure transmission of record traffic (narrative and graphic) between tactical operations centers (TOCs), LOS multichannel facilities to tie in high-volume traffic requirements between division main, rear, and division support (DISCOM), and a general support reserve augmentation capability (TACSAT) located at the corps level.

The command operations company (1) provides the telecommunications center capability for Division Main and the Division Tactical Operations Center (DTOC); (2) meets telecommunications center requirements for DISCOM and provides multichannel communications for Division Main, DISCOM, and rear; and (3) provides 30-line SB-3865s, access units, and record traffic capability at the All-Source Analysis Center (ASAC), rear CP, airfield, Alt, and TAC. Both Alt and TAC have single-channel TACSAT record traffic capability as an added requirement.

The forward operations company (1) deploys six mobile subscriber centrals throughout the division area; (2) provides 30-line SB-3865s, access units, and teletype; (3) provides a multichannel TACSAT access section using three AN/TSC-93 terminals for augmentation as required; and (4) provides a wire and cable platoon capability.

Most of the major equipment will not be available until 1986 or later. Therefore, the implementation will not be discussed in detail at this time. In addition, the Army is going through a division restructuring that would impact any scenario or deployment projected at this time.

4.8 CORPS COMMUNICATIONS

The corps communications support system is based on the INTACS concept and uses the INTACS system architecture to achieve the objective system and meet the requirements of the future in all four phases of corps communications:

Phase I—Restructure of the corps and deployment of improved INTACS equipment

Phase II—Integration of new equipment

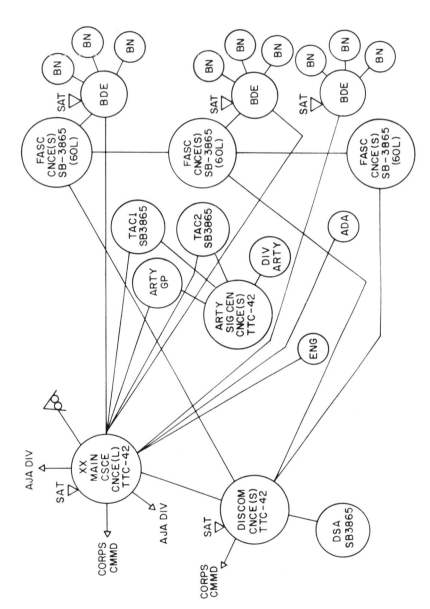

Figure 4.3 Division Multichannel System Digital System (FM 11-50 Doctrine).

Phase III—Deployment of corps digital multichannel switching systems
Phase IV—Objective system application for corps elements

The architecture utilizes the design specifications and operational characteristics of the current equipment and the equipment being developed under the improved ATACS, TRI-TAC, SINCGARS, MSE, and SATCOM programs to achieve the operational capabilities required to successfully perform the Army's mission.

The architecture's design and flexibility is such that it can adapt to changing requirements and incorporate equipment developed in the future. The architecture presents criteria for connecting today's tactical communications systems into the objective system and maintaining the system. The system of the future is based on state-of-the-art equipment and a new organizational structure that provides the base design for a total telecommunications system.

The corps objective system consists of 12 nodes for the entire corps area to include displacement capability. The signal organization to support the system is composed of four battalions of three companies each (Figure 4.4). The battalion has a Communications System Control Element for management and control, a store-and-forward module, AN/TYC-39, and antenna towers. All other communications equipment is organic to the companies.

It is envisioned that ten deployed nodes can adequately support the corps deployment, leaving two nodes available for displacement. The connectivity between nodes consists of AN/TRC-138 (relayed as required) high-capacity systems organic to the area battalions/companies.

Each node has the capability of establishing six low/medium-capacity radio extension links to units in the vicinity of the node plus two coaxial cable links. There are three CNCEs (small) and three 75-line unit-level switchboards available for use on those extensions. That switching capacity gives the system the capability of additional link connections to other mininodes through the use of the spare radio in the AN/TRC-174. Flexibility in routing in the event of node failure or destruction is greatly improved.

Within the perimeter of each node there is a complex consisting of one 300-line AN/TTC-39(V) and an appropriate quantity of high-capacity and low-capacity radio terminal equipment to properly terminate all radio links. Every third node has a store-and-forward module AN/TYC-39 for record traffic switching and a TACSAT terminal.

4.9 NAVAL COMMUNICATIONS

Naval communications are used, among other things, to link naval commands to the operating forces ashore, at sea, and in the air. Naval communications carry command and control information to fleet units and status and feedback information from the fleet units and forces to commands. Communications also distribute surveillance-sensor and intelligence data both to command and to the forces at sea and extend laterally among the operating force units, allowing them

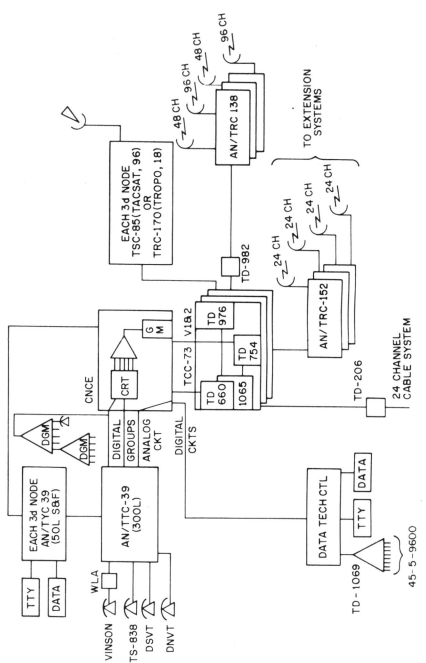

Figure 4.4 Corps area node—Transition hybrid early.

to keep one another informed and coordinated and mutually supportive during the performance of their assigned tasks and missions. Communications tie fire control instructions and update information to weapons and tie operational units to the Navy logistics structure, enabling these support forces to respond rapidly to the needs of the fleet.

4.9.1 The Naval Telecommunications System (NTS)

The communications that make up the Naval Telecommunications System (NTS) take many forms. Beyond the Line-of-Sight (BLOS), long-range communication is needed to send long-range broadcasts from major operational and administrative units and to provide a one-way path for promulgating command decisions. Long-range ship-to-shore links provide for the necessary acknowledgments, status reports, and information exchanges needed to coordinate fleet operations. Extended Line-of-Sight (ELOS) and Line-of-Sight (LOS) communications maintain short-range and medium-range ship-to-ship, ship-to-air, air-to-ship, and air-to-air links for the information exchange needed to coordinate the movements of individual fleet units and formations. Computerized combat direction systems in ships and aircraft exchange vital sensor, fire control, and electronic warfare information via a number of high-speed data links. Tactical radio-telephone systems back up these high-speed links with voice communications, providing an extra measure of flexibility at the tactical level.

Today these communications links remain more vulnerable to jamming and exploitation than ever before. They are being upgraded in both quality (more survivable, more jam resistant, more secure) and quantity (more capacity to meet the growing communications and information transfer requirements of the Navy). There are programs underway to achieve these requirements (e.g., the HF Improvement Program, the UHF Anti-Jam Line-of-Sight (AJ LOS), the Joint Tactical Distribution Information System (JTIDS), and Advanced SATCOM). These programs go a long way, but new, more capable communications systems and links need to be developed if future quality and quantity requirements are to be fully met. Not all of these features and their associated costs are justifiable for all the users of Naval communications. Some user communications requirements (e.g., routine logistics, maintenance, and personnel traffic) cannot justify the above levels of protection and service. The Naval telecommunications architecture provides for three levels of communications service: minimum essential communications (MEC), unimpaired tactical effectiveness (UTE), and normal sustained operation (NSO). It is the required minimum essential communications capability that must be afforded the highest level of communications service outlined above. And it is the communications traffic necessary to effectively support some minimal level of command and control of the forces at sea that represents this minimum essential communication capacity.

The Naval Telecommunications System (NTS) operates and maintains a worldwide network of Naval communications stations (NAVCOMSTAs) for the Defense Communications System (DCS)). They are controlled via four Naval communications area master stations (NAVCAMS), located in Norfolk, Naples, Honolulu, and Guam. Those communications areas correspond roughly to the areas of responsibility of the theater commands: CINCLANTFLT (NAVCAMS Norfolk), CINCUSNAVEUR (NAVCAMS Naples), and CINCPACFLT (NAV-CAMS Honolulu and Guam). The NAVCOMSTAs transmit the Fleetbroadcast. The primary Fleetbroadcast is provided by satellite but it is backed up with an HF shore-to-ship broadcast.

As discussed earlier, The Defense Communications System (DCS) provides the basic communications for worldwide command and control of U.S. military forces. The current DCS is composed of four principal systems: the Automatic Digital Network, AUTODIN, a digital message-switching communications system; the Automatic Voice Network, AUTOVON, an analog voice circuit switch communications system; the Automatic Secure Voice Communications Network, AUTOSEVOCOM, a narrowband and wideband secure voice link to about 1,500 AUTOVON subscribers; the Defense Satellite Communications System (DSCS II), an SHF communications system comprised of active geostationary satellites that operates in support of the World Wide Military Command and Control System (WWMCCS); and numerous dedicated circuits. AUTOVON and AU-TODIN are currently planned to be upgraded and to transition into the Defense Switched Network (DSN) and the Defense Data Network (DDN), respectively.

4.9.2 Satellite Communications

The Fleetbroadcast satellite communications component of the FLTSAT and GAPFILLER, is currently a 16-channel broadcast, SHF uplink and UHF downlink. The satellite receiver system is the AN/SSR-1A. Its demux, the UCC-1, is planned to be upgraded in 1985/86. The NAVCOMSTAs also serve as the link between the DCS and the ships at sea. There are several additional links to the fleet beside the Fleetbroadcast channels that also reside on the FLTSAT and the GAPFILLER. The Common User Digital Information Exchange System, CUDIXS, is a GENSER (general service) message-handling and processing system to and from ships. Messages are processed on ship by the NAVMACS system. The shore-based portion of the message-processing system is controlled by the Naval Computer Message Processing and Relay System (NAVCOM-PARS) and the Local Data Message System (LDMS). Precedence determines the relative importance, or priority, of messages, and thus the order in which they will be transmitted. The precedences in ascending order are Routine, Priority, Immediate, and Flash.

The Submarine SATCOM Information Exchange System (SSIXS) provides submarine broadcast messages in unformatted text and targeting information in formatted computer-readable, character-oriented NATO standard message format called RAINFORM format.

The Officer in Tactical Command Information Exchange System (OTCIXS) is the battlegroup command and control channel. Data at 2400 bps (RAINFORM format), teletype at 75 baud, and narrowband Secure Voice at 2400 baud can all be sent over this channel.

The Tactical Intelligence Network (TACINTEL) is for the exchange of special intelligence (SI) traffic. It employs the AN/USQ-64(V)5 and carries both RAINFORM formatted messages and unformatted narrative text.

The narrowband secure voice channel uses the CW-3333 vocoder and the ON-143 interconnecting group. It is used to support the Hi Command Net linking the Fleet Command, the Commanders-in-Chief (CINCS), and the battlegroup commanders. It also contains the common user narrowband secure voice, FLT TAC.

The FLTSAT and the GAPFILLER satellites are UHF. The shipboard system uses the AN/WSC-3(V)3 for UHF satellite and the AN/WSC3(V)8 for line-of-sight (LOS) UHF. Several improvements are planned in this frequency band. They include the use of the advanced narrowband digital voice terminal (ANDVT, the CV 3591) to solve the interoperability problems between wideband and narrowband secure voice and the UHF demand-assigned multiple-access multiplex (DAMA, the TD-1271B/U) to increase the number of channels that may be supported over the UHF satellite. One 25 KHz satellite channel from the WSC-3(V)3 is divided into four 2400 baud channels (one 2400 baud channel may in turn be divided into six low-level teletype channels at 75 baud each). There is also a joint service EHF satellite (MILSTAR) planned that would replace current UHF FLTSAT and GAPSAT satellite channels with more jam-resistant EHF channels. (This is discussed in more detail in Chapter 5.) The Navy shipboard EHF terminal for this system is the AN/USC-38, scheduled for initial operational capability (IOC) in the first half of the 1990s.

In addition to these FLTSAT channels, the DSCS II relays communications in the SHF band to, among others, the fleet command centers ashore and to the task force command centers aboard the major flagships. Because of the size and complexity of current SHF receiving terminals, few Navy ships, other than flagships, are able to communicate directly via these satellites. There currently is an upgrade for the DSCS II, called the DSCS III; it will operate in the SHF band and possibly, in the EHF band.

Virtually all seagoing U.S. Navy ships communicate with each other in a variety of frequency bands. Most ships copy the Fleetbroadcast over UHF via teletypewriter at over 100 words per minute. There is also, as discussed above, a common fleet tactical UHF frequency that operates line of sight. UHF frequencies are used routinely between ships and aircraft operating in company as long as they are within line of sight of one another. For longer-range ELOS communications between ships and aircraft, HF communications can be established, usually ground wave within a force and sky wave for ship-to-shore. These tactical HF or UHF links are used both for voice and Navy Tactical Data System (NTDS) equipped ships and aircraft digital links.

4.9.3   The Naval Tactical Data System (NTDS)

The NTDS, the principal means by which Navy ships and aircraft exchange tactical data, provides the officer in tactical command with real-time strategic and tactical information. NTDS has three major links:

- Link 11 is a two-way, real-time, encrypted, but not currently jam-resistant, data link that operates in both the HF and the UHF frequency ranges. Link 11-equipped platforms exchange tactical sensor data, as well as weapon deployment and engagement status data. The link protocol is a time-ordered synchronous, bit-oriented message format operating at 2400 bps or greater.
- Link 14 is one-way UHF or HF from NTDS-equipped ships to other surface ships. Using Link 14, Link 11-equipped ships can transmit encrypted data to ships lacking the complex NTDS processing equipment. That raw data is transmitted at 75 baud and is transcribed in character-oriented text, in a form suitable for presentation to the onboard commander.
- Link 4A is a one-way, time-division multiplex link used to control interceptor aircraft. NTDS surface ships and Aircraft Tactical Data System (ATDS) aircraft can control a suitably equipped interceptor aircraft via Link 4A, as well as provide vector information for the pilot to follow manually.

A good many of these existing tactical links, both data and voice, are not jam resistant. The Joint Tactical Information Distribution System (JTIDS) is planned to replace most of the existing voice and data tactical links with a higher capacity, jam-resistant wideband link. (A more detailed description is provided in Appendix A.)

Manual Morse is sometimes still used, as backup, over HF. Encryption is used for some voice links and for most data links. The ultimate goal is to encrypt all links; affordability versus perishability of the information is one of the key issues here. Voice links are switched from the ship's communications center to remote handsets throughout the ship, e.g., the bridge and the combat information center.

4.9.4   Submarines

Communications with submarines represent a particularly unique situation. Because of the covert nature of submarine operations, submarines normally communicate in a "receive only" mode. There is a submarine broadcast in both the VLF band, via special very high power transmitters and via the SSIXS FLTSAT channel, just as there is a Fleetbroadcast for the surface fleet. Submarines can also participate in other SATCOM links and acoustic communications for coordinated ASW operations and to receive targeting data. In order to receive the VLF broadcast while submerged, the submarine must trail an antenna wire on or near the surface of the water or tow a communications buoy slightly beneath the surface. When communicating in other standard Navy frequency bands, submarines must raise an antenna above the water surface. These communications systems require the submarine to operate relatively near the surface and also impose constraints on the submarine's speed and maneuverability. This exposure increases the submarine's risk of detection.

Because of its need to be in constant communication with the National Command Authority (NCA) for strategic firing orders but yet to remain covert and undetectable, the strategic ballistic missile nuclear-powered submarine (SSBN) has been equipped with a variety of exotic communications systems. The deployed SSBN force is served by the Minimum Essential Emergency Communications Network (MEECN), which links the strategic retaliatory forces (Strategic Air Command bombers, missiles, and Navy SSBNs) to the NCA via a highly redundant communications system. In addition to receiving the VLF broadcast transmitted directly by shore station, patrolling SSBNs can also receive VLF communications relayed by specially configured Navy TACAMO aircraft. TACAMO aircraft receive several redundant uplinks from ground- and air-based elements of the NCA in the VLF through UHF bands. They relay these communications to patrolling SSBNs at VLF frequencies using 10,000 to 35,000 foot long trailing-wire antennas.

The Navy is developing ELF and laser communications systems to reduce the need for U.S. submarines to operate near the ocean surface. ELF transmission penetrates the water as much as 20 times deeper than VLF transmissions, allowing submarines to receive communications at greater operating depths. Like VLF, ELF transmissions are basically immune to atmospheric disturbances, either natural or caused by nuclear bursts. The major disadvantage of ELF is the very low data rate it supports. Therefore, the ELF system would be useful primarily as a call-up system. It would enable shore-based authorities to execute preplanned emergency actions or to alert submarines that an important message was being sent through normal channels. In this way, the submarine could avoid operating near the surface unless called up to receive a high-volume message.

An ELF system requires a transmitter antenna that extends for hundreds of miles. One approach is to build a buried grid. For the antenna grid to propagate the ELF signal most effectively, it should be located atop geologically "old bedrock." This limits the number of possible sites for ELF. (Laser communications is discussed in more detail in Chapter 8.)

4.10   AIR FORCE COMMUNICATIONS

Air Force communications requirements in support of land, sea, and air battles for strategic and tactical applications are unique in some aspects yet compatible with those of the other services in certain areas. This section provides an overview of some of the critical Air Force systems.

Once again, the goal is to improve the security, jam resistance, connectivity, and capacity of present systems. Of particular interest to the Air Force are data and voice communications between aircraft, from aircraft to ground, and from ground to ground that provide C^3 and sensor/weapon data links. The important technological developments that help meet these requirements are discussed in Chapter 8.

During the 1980s, many improvements will be made in the Air Force's ability to communicate in a battlefield environment. Programs like JTIDS, TRI-TAC, and the ground mobile forces satellite communications terminals will improve the security, jam resistance, connectivity, and capacity of Air Force tactical communications. However, some important problem areas will remain even after these programs have been implemented. Of particular concern are the survivability problems associated with the capability of the equipment to resist the accelerating electronic warfare and physical threats of a determined enemy.

The goal of the Air Force communications system planning activities is to reduce or eliminate the remaining problem areas in the 1990s. The basic context and overall C^3 architecture for the communications system planning and system engineering activities are provided by the Master Plan [1] for the Tactical Air Forces Integrated Information System (TAFIIS). That plan was generated jointly by two Air Force organizations: the Tactical Air Forces Interoperability Group and the Deputy for Development Plans of the Electronic Systems Division. Much of the material in this chapter is directly related to the TAFIIS Master Plan and has been strongly influenced by it. In particular, the plan's recommendations for a dispersed, distributed, survivable C^3 system for the 1990s have a major impact on communications tradeoff analyses and the conclusions and recommendations that result from such analyses.

The description of current Air Force tactical communication systems and equipment in this section is organized into two categories: aircraft systems and ground-to-ground communications systems.

4.10.1 Aircraft Systems

For tactical operations, voice communications are required among tactical aircraft and between tactical aircraft and ground control units. UHF radios in the 225-400 MHz band provide most of the Air Force's tactical voice communications capability; all Air Force tactical aircraft carry at least one UHF/AM radio. In addition, aircraft involved in close air-support missions carry VHF/FM radios. Until recently, many different types of UHF and VHF radios were used in tactical aircraft. Now, older radios are being replaced by newer, standardized, solid-state radios.

The AN/ARC-164 has replaced most older UHF radios and is now the principal UHF radio in tactical Air Force aircraft. Operating in the 225-400 MHz band, with 7,000 channels available at a channel spacing of 25 kHz, its power output is 10 W with a 30 W option. It provides half-duplex (push-to-talk) voice communications, either secure or nonsecure and has no special provisions for ECCM.

Another UHF radio, the AN/ARC-171, is employed in a much smaller number of tactical aircraft (e.g., AWACS). Its characteristics are output power up to 30 W for AM voice, up to 100 W for FM voice, data, and satellite communications; compatibility with some DoD digital data links; and nuclear effects protection.

The principal VHF radio is the AN/ARC-186. Its power output is 10 W in the AM mode and 15 W in the FM mode. The channel spacing will be 25 kHz, and the equipment is compatible with COMSEC equipment.

HF radios are used only in larger aircraft (e.g., AWACS) and in reconnaissance aircraft at long ranges. As an example, the AN/ARC-165 is used for long-range air traffic control and for command and control purposes. Its output power is 1000 W single sideband (SSB) and 250 W independent sideband (ISB).

Airborne equipment for data communication is generally lacking in tactical aircraft today. Important exceptions are some airborne sensor platforms (e.g., AWACS), reconnaissance aircraft, and air-to-surface missiles that use special-purpose data links often lacking in ECCM capability.

4.10.2   Ground-to-Ground Communications Systems

Most of the communications systems and equipment currently employed for Air Force tactical ground-to-ground communications are based on the technology of the early 1960s. As such they tend to be analog, nonsecure, slow, labor-intensive, and not well-suited to handling the digital traffic expected in the 1980s and 1990s. Many different equipment types are in use by the Tactical Air Forces; some of the most important of those are described briefly below, with emphasis on deployable assets.

Multichannel transmission is provided primarily by analog nonsecure troposcatter equipment. The primary example of this type of equipment is the AN/TRC-97A. It operates in the 4.4-5.0 GHz band and is based on analog, voice-bandwidth FM/FDM operation. Eight-foot parabolic antennas are employed, with dual space diversity. The maximum transmitter output is 1 kW; however, when the TRC-97A is used in the optional line-of-sight mode, the power is reduced to 1 W and a smaller horn antenna is employed. The TRC-97A can handle voice (up to 24 channels), teletypewriter inputs, and digital data (theoretically, up to a rate of 2400 bits/s per voice channel). Its range is limited to less than 160 km, especially when data BER requirements must be met.

HF equipment is also employed for ground-to-ground transmission in contingency operations. Principal applications are for initial setup of ground-to-ground links, backup for troposcatter links, and connections among ground-based forward air controllers, operations centers, and Army units.

The basic telephones employed today are analog nonsecure instruments. Some limited provisions for secure voice communications are included among the deployable communications assets for special circuits and for providing access to the worldwide Automatic Secure Voice Communications System (AUTO-SEVOCOM). All of the security equipment requires special interface considerations.

Teletypewriter equipment is employed extensively in the tactical Air Force centers at the present time; typical is the Teletype Model 28. Some limited use is also made of the Teletype Model 40 teletypewriter, but it does not meet military specifications for tactical use. In general, the tactical teletypewriters do not have capabilities for text editing and text memory (except via paper tape).

For digital data, the tactical communications assets include a number of interfaces and modems to transform data streams up to 1200 bits/s into signals compatible with 3 kHz channels. A limited capability for 2400 bits/s transmission over 3 kHz channels is also available. Facsimile terminals are also used in Air Force tactical applications, but not widely.

Today's tactical circuit switches are oriented toward analog, voice-bandwidth operation. The principal automatic, deployable circuit switch is the AN/TTC-30. It is truck-transportable and automatically establishes and controls connections within the Air Force site at which it is located and between sites. Some of its service features are conferencing, preemption (two levels), direct-access capability ("hot-line" or dedicated), and automatic alternate routing.

With regard to message switching, the only current tactical message switch/ relay included among the deployable Air Force communications assets is the AN/TGC-26. This switch provides only torn-tape relay operation, which is inherently labor-intensive, slow, and subject to major delays in high-traffic situations. It operates at 75 baud (100 words/min) and provides up to 24 full-duplex teletypewriter circuits. A multiple-address handling capability is included, to assist in the preparation of messages intended for many recipients.

At present, no network control facilities exist within the deployable Air Force communications assets. The only control functions performed are done on a nodal or a link basis, not on a networkwide basis; these functions are referred to as "technical control." There are three basic technical control functions: interfacing, control, and reporting. In the interfacing function, the control facilities provide interfaces between terminal equipment and transmission equipment and between different transmission media. The control functions include assessment of equipment and circuits, fault isolation, coordination for reconstituting links, rerouting of circuits to avoid service degradation, reconnection of circuits on the basis of user precedence levels, and coordination for initial path connections. The current tactical technical control facilities are all analog and manual.

4.10.3   Future Air Force Systems

Many of the planned Air Force systems are similar to those the Army is procuring through the TRI-TAC program. However, some are peculiar to the Air Force. This section concentrates on the following areas: aircraft voice, aircraft data, ground-to-ground communications, and data-link development.

4.10.3.1   Aircraft Voice

Three major programs will have a significant impact on aircraft voice communications in the 1980s: the HAVE QUICK and enhanced JTIDS (EJS) systems, which will affect UHF radios, and the SINCGARS program, which will affect the Air Force's VHF radios.

The objective of HAVE QUICK and EJS [2] is increased voice capability for tactical aircraft in a battlefield environment. HAVE QUICK will provide interim, or near-term, modifications to the present UHF/AM voice radios to achieve a degree of jam resistance for air-to-air and air/ground/air voice communications. The program is aimed at near-term electronic-warfare threats.

Enhanced JTIDS is a longer-range program that is aimed at future, high-threat, electronic-warfare environments. In the mid 1980s, it should take over from HAVE QUICK and provide jam resistance, voice conferencing, and interface with secure voice devices, in that order of importance [3]. (The Air Force users have expressed a desire for an air-to-air voice-conferencing capability. If it is not practical to provide this capability with full security, they prefer the conferencing capability over the full security capability in some dynamic battle situations because it is transient, short-term security or privacy that is the principal requirement, rather than full security.) The EJS radios will retain the major features of present tactical UHF/AM radios and will use direct-sequence, spread-spectrum, and adaptive-array techniques for nulling jammers. The final design will be selected on the basis of parallel engineering-development models that will be tested and evaluated on a competitive basis.

In the VHF band the key program is the Army SINCGARS (Single Channel Ground and Airborne Radio Subsystem) effort. In the late 1980s, Air Force tactical aircraft must be able to interoperate with the ground-based Army SINC-GARS secure, antijam radios for certain tactical operations. Those radios are to be compatible with the present VHF/FM radios in the 30–88 MHz band, including the Air Force's ARC-186. Voice will be digitized at 16 kbits/s and security will be provided via equipment that is compatible with TRI-TAC's COMSEC equipment. Data transmissions are to be interoperable with tactical Army data systems, with variable data rate capabilities up to 16 kbits/s. SINCGARS ECCM add-on features, which will reduce vulnerability to enemy interception, direction finding, and jamming, will include frequency hopping and provisions for reducing RF power output. In addition, the possibility of employing a steerable-null antenna is being explored. The Air Force is investigating several alternatives for providing interoperability with ECCM-equipped SINCGARS radios, including possible modifications to or replacements for the ARC-186.

In the HF band, no acquisition programs currently exist for new radios for tactical aircraft. However, new HF radios are planned for strategic aircraft.

4.10.3.2 Aircraft Data

A major improvement in the data-communications capability of tactical aircraft will be introduced by means of the Joint Tactical Information Distribution System (JTIDS). (Because of the importance of JTIDS in all three services communications plans, an extended description of the system is provided in Appendix A.) JTIDS is a time-division multiple-access (TDMA) system that is being developed to provide secure, jam-resistant information transfer among tactical aircraft and ground C^2 elements. Most of the information transferred is expected to be digital data, but a limited amount of digitized voice can also be handled.

JTIDS can also perform position location and identification functions, which will be carried out in the 950-1215 MHz band.

The digital data carried via JTIDS will include aircraft tracks, position and status reports, ground-to-air missile threats to Air Force aircraft, target assignments, and mission results. In its initial implementation, JTIDS will be incorporated into the AWACS aircraft and in ground-based interface terminals that will allow air-surveillance data to be exchanged with C^2 centers. Follow-on implementation plans envision JTIDS terminals being introduced into air-superiority fighters and then into air-ground mission aircraft and control facilities.

The intention of the basic JTIDS design philosophy is to allow an information user to select needed information from an "information pool." A TDMA "bus" is employed to implement this philosophy. As indicated in Figure 4.5, the various elements involved in a tactical operation are assigned time slots for their transmissions; when an element's turn to transmit arrives, it transmits information that may be of interest to other members of the specific "net" on which it is transmitting. Depending upon the geographic distribution of terminals, about 20 JTIDS nets can be operated simultaneously in the same geographical area (without degrading AJ characteristics) by using code-division techniques. Each net has an information data rate of 57.6 kbits/s; this rate is reduced to 28.3 kbits/s if the forward-error-correction option is used. A synchronization preamble is transmitted with each message and a "guard time" permits each message to propagate through the geographic coverage area before the next time slot begins.

Three classes of terminals have been considered for the JTIDS program [3]. Class I terminals are intended for large aircraft (e.g., AWACS), surface ships, and interface facilities that link JTIDS to ground-based networks. Class II ter-

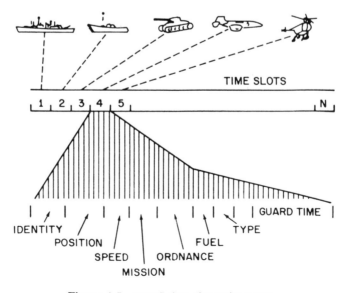

Figure 4.5 JTIDS time slot assignments.

minals are intended for fighter-class aircraft and for ships with volume con-straints. The volume of the Class II terminals will be about 2 ft^3, which will be smaller than the Class I terminals; however, similar functions will be performed. The Class III terminal is envisioned for use by forward air controllers, small remotely piloted vehicles (RPVs), some theater missiles, and selected Army units. The Class I and II terminals have 1000- and 200-W power amplifiers, respectively. Other specific characteristics of these terminals are identified in Appendix A and in [3].

Other features of the JTIDS design are: both formatted and unformatted mes-sages can be transmitted; any aircraft terminal can act as a relay; and messages can be directed to a single user if desired.

Two possible improvements to the basic TDMA approach are being investi-gated: advanced TDMA and distributed TDMA. In ATDMA, additional data can be packed into the basic TDMA time slots and the time slots can be divided into halves to provide more opportunities for users. In DTDMA, data symbols from different messages are interleaved, in contrast to the burst message structure of the basic JTIDS TDMA. Both ATDMA and DTDMA will have provisions for accommodating terminals employing the basic TDMA approach.

4.10.3.3 Ground-to-Ground Communications

Descriptions of new ground-to-ground communication equipment are presented in this subsection. However, the descriptions of the TRI-TAC equipment are abbreviated. Typical Air Force interconnections of TRI-TAC equipment are depicted in Figure 4.6.

Transmission Equipment

In the 1980s, a family of new troposcatter terminals will be introduced into the Air Force's tactical inventory. That equipment, developed under the TRI-TAC program, is designated as the AN/TRC-170. It will transmit digital signals and the largest version will feature—in comparison with today's AN/TRC-97A—larger antennas, quadruple diversity, higher power (up to 6.6 kW per trans-mitter), a lower receiver noise figure (3 dB), longer range (up to 320 km), more voice channels, higher data rates (up to 2 Mbits/s), security, ECCM provisions, and time-division multiplexing rather than frequency-division multiplexing. Ad-ditional details concerning the TRC-170 can be found in Appendix B or in [4] and [5].

An important aspect of the TRC-170 is that it is completely compatible with the other equipment being developed under the TRI-TAC program. Thus, it offers the possibility of significantly reducing the interservice interoperability problems that exist with today's troposcatter equipment. With regard to satellite communications, a significant improvement in mobility and setup time for mul-tichannel links will be introduced into the Air Force's tactical capabilities in the 1980s via the GMF terminals. Many of the details of the GMF program can be found in [6], but since the Air Force's GMF plans differ somewhat from the

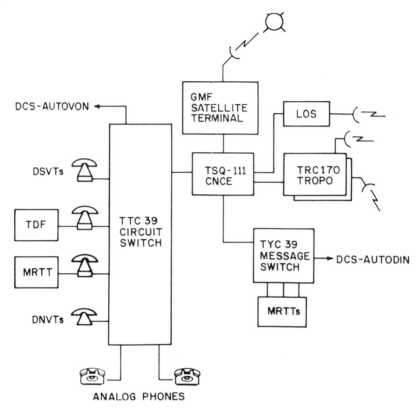

Figure 4.6   TRI-TAC and GMF equipment at a representative Air Force node.

Army's, additional details of the unique aspects of the Air Force approach are presented in this subsection.

The Air Force portion of the GMF program will be implemented in three phases. In the first phase, an initial capability will be provided using GMF ground terminals designated as the AN/TSC-94 and the AN/TSC-100.

The TSC-94 can be used to establish a single link between two ground locations. It provides 12 voice-frequency channels, with voice digitization accomplished via the 48-kbit/s PCM approach utilized by the Army. The terminal's output bit stream is provided with traffic flow security prior to transmission. Some significant characteristics of the TSC-94 are truck-transportability, operation in the 7-8 GHz band, an 8-ft antenna, 500 W transmitted power, receiving system noise-temperature of 300 K, and an overall G/T of 17.7 dB/K.

The TSC-100 terminal is larger than the TSC-94 and can act as a "hub" to establish up to four links to four different locations. The TSC-100 uses a 1000 W transmitter and can employ a 20-ft antenna, in addition to the 8-ft antenna utilized by the TSC-94, for two-antenna operation that allows dual-satellite/dual-transponder usage. The receiving system of the TSC-100 is similar to that of the TSC-94.

The second phase of the Air Force's GMF plans will add ECCM provisions, the ability to form a "mesh" connectivity, and interoperability with TRI-TAC equipment and NATO. The new terminals will be designated as the Non-Nodal Terminal (NNT, TSC-94A) and the Nodal Mesh Terminal (NMT, TSC-100A). The utilization of these terminals in a representative Air Force deployment is shown in Figure 4.7.

During normal operation, the NNT will handle a full TRI-TAC group (up to 512 kbits/s) plus up to 24 individual full-duplex subscribers (to a maximum of an additional 512 kbits/s). The individual subscriber end instruments may be any combination of voice, data, or teletypewriter traffic. Under jammed conditions, up to 24 individual subscribers can still be served, provided that the combined total bit rate does not exceed 32 kbits/s. Depending on the specific threat scenario, graceful degradation from 32-kbits/s capability can be accomplished. At baseband, the NMT differs from the NNT only in the quantity of circuits it can handle. The NMT will accept up to six TRI-TAC groups and/or up to 72 individual subscriber accesses. During jammed conditions, the communications capability will be limited to a maximum of 32 kbits/s per link.

The radio-frequency (RF) portion of the NNT is similar to the TSC-94 terminal. It will, however, have the capability to combine the output of two high-power amplifiers to double the power output when necessary. The RF portion of the NMT is similar to that of the TSC-100, and, like the TSC-100, it has the additional flexibility of two-antenna operation (8 ft and 20 ft), to allow networking with

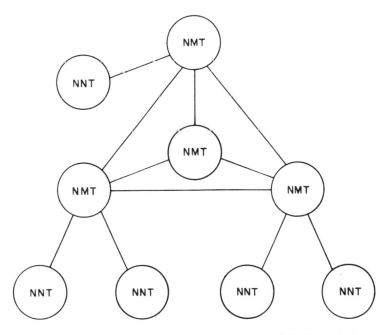

Figure 4.7 Representative Air Force deployment of GMF terminals.

two satellite transponders for inter- and intratheater connectivity. Both the NNT and NMT meet tactical mobility and transportability criteria.

All of the GMF terminals described above will operate initially with the DSCS II satellites and later with the DSCS III satellites, as described in [7] and [8]. During unstressed operation, they will employ frequency-division multiple access (FDMA), sending multiple carriers through the satellite transponder. Spread-spectrum multiple access (SSMA) will be employed for operation in a jamming environment.

A third phase is also envisioned in the Air Force's plans for GMF satellite terminals. While the details of this phase have yet to be finalized, it will probably involve the GMF program plans for its Multichannel Objective System (MCOS) [6] and its Single Channel Objective Tactical Terminal (SCOTT). The MCOS and SCOTT will employ demand-assigned multiple-access techniques and may use an EHF satellite with onboard processing in the 1990s.

With regard to line-of-sight links, a new radio is envisioned under the TRI-TAC program; the current designation for the radio is the Short Range Wide Band Radio (SRWBR). It will be completely compatible with the TRI-TAC family of equipment and will operate at various rates up to approximately 20 Mbits/s at normal line-of-sight ranges (which depend upon the terrain).

Other possibilities for short ground-to-ground links are (1) fiber optic cables, (2) coaxial cables, and (3) the use of the TRC-170 in a line-of-sight mode. A future possibility is the use of EHF or optical (e.g., laser visible or ultraviolet) line-of-sight links.

End Instruments

A number of new end instruments that will be important additions to the Air Force's tactical communications inventory are currently in development under the TRI-TAC program: the Digital Secure Voice Terminal (DSVT), the Digital Nonsecure Voice Terminal (DNVT), the Advanced Narrowband Digital Voice Terminal (ANDVT), the Modular Record Traffic Terminal (MRTT) family, and the Tactical Digital Facsimile (TDF).

The Digital Secure Voice Terminal is a full-duplex telephone that provides secure access to the TRI-TAC circuit-switched network. Voice is digitized at a bit rate of 32 kbits/s (or 16 kbits/s) and encryption is provided by equipment compatible with the family of COMSEC equipment being developed for TRI-TAC.

The Digital Nonsecure Voice Terminal is a full-duplex telephone that digitizes voice at 32 kbits/s (or 16 kbits/s). It is compatible with TRI-TAC equipment and can interoperate with a secure telephone if the proper encryption equipment is provided at a TRI-TAC switch.

The Advanced Narrowband Digital Voice Terminal will provide half-duplex, encrypted voice communications capability at a bit rate of 2400 bits/s.

The MRTT family of secure, ruggedized record-traffic equipment will provide facilities to compose, edit, transmit, and receive record traffic.

The Tactical Digital Facsimile (TDF) equipment will provide a wide range of facsimile capabilities for tactical use. With this flexibility, the traditional role played by facsimile equipment is expected to expand to include some of the record communications that have previously been provided by teletypewriters. The Air Force plans to use the TDF as the facsimile unit in its Intratheater Imagery Transmission System (IITS).

Switches

The TRI-TAC program will introduce several new circuit switches and message switches that will eliminate many of the current problems with tactical Air Force switches. All of the switches will be truck-transportable or man-transportable.

The automatic circuit switches being developed are designated the AN/TTC-39, AN/TTC-42, and SB-3865; the latter two, known as Unit Level Switchboards, are significantly smaller than the TTC-39. These switches will provide service for both digital communications systems of the future, including effecting connections between certain analog and digital end instruments. The TTC-39 is expandable to a maximum of approximately 600 terminations and may be further expanded up to a maximum of 2,400 terminations by collocating switches. It will handle voice, record traffic, data, facsimile, and imagery.

The TTC-42 will be available in 75- and 150-line versions. The basic SB-3865 has 30 lines and those basic units can be cascaded into 60- and 90-line versions.

TRI-TAC message-switching development is following a pattern similar to that for circuit switches; a larger switch, the AN/TYC-39, the AN/TYC-11, and the AN/GYC-7 are being used. All of these switches are automatic.

The TYC-39 will be a digital, secure, store-and-forward message switch that will replace current Air Force torn-tape relays. It will have up to 50 terminations.

System Control

System control for future Air Force tactical communications will be based primarily on the family of TRI-TAC developments known as the Tactical Communications Control Facilities (TCCF). When fully implemented, the TCCF will help decision makers manage networks and equipment by collecting, processing, and displaying data for operating personnel. As currently envisioned, four hierarchical levels will eventually be developed under the TRI-TAC program [4]:

1. Communications System Planning Element (CSPE)
2. Communications System Control Element (CSCE)
3. Communications Nodal Control Element (CNCE)
4. Communications Equipment Support Element (CESE)

4.10.3.4 Communications System Planning Element (CSPE)

The CSPE is the highest level within the management and control hierarchy. This element will be management oriented. Functionally, the CSPE staff will provide broad system-planning and system-engineering services relative to a designated communications system or network. Based on operational plans or orders provided to the CSPE by the commander's tactical operations element and statistical data gathered by the CSPE, communications system-planning information will be developed and many staff functions will be performed prior to deployment. These staff functions will normally include preparation of estimates for long-range system requirements, contingency planning, determination of initial switched-network and transmission-system configuration, assessment and allocation of available resources, specification of locations of interfaces and methods of interfacing with the control elements of DCS allied and other communications systems, long-term management of COMSEC resources, and continuity of technical management and control. Initially, CSPE functions will not be automated, but will use the data base available from associated subordinate CSCEs or CNCEs. As expertise in the area of computer-assisted management and control of communications resources develops, many CSPE functions will become candidates for automation.

4.10.3.5 Communications System Control Element (CSCE)

The CSCE provides the primary near real-time management and control of the subordinate communications facilities located at the nodal system as directed by the CSPE staff. A node consists nominally of one or more circuit switches, a communications nodal control element (CNCE), network extension facilities, and associated transmission equipment. It is anticipated that the CSCE will normally provide primary control and management of up to eight nodes and will function as a backup to another CSCE for the control and management of up to an additional eight nodes. In the event of failure of a primary CSCE, the associated backup CSCE will, therefore, be capable of assuming primary control and management of the additional nodes with a corresponding reduction in management capability.

The CSCE will be an automated facility maintaining a current data base on the status of the entire subordinate communications network. That data base will include traffic data; system/equipment performance data; equipment restoral status; grade of service; circuit outages and backlog situations; network connectivity to include internodal routing/connectivity, restoration priorities, channel paths, and dedicated circuit and message switch directories for the entire network; and a list of COMSEC resources. The data base enables the CSCE to provide, as a minimum, near real-time control of the following aspects of the communications system.

- Traffic control of the circuit and message switch network by directing appropriate implementation of alternate routes, trunk barring, line load control, and trunk/link directionalization
- Transmission systems routing control of the trunk transmission network by directing reallocation/allocation of transmission channels between nodes based on restoration priorities and/or traffic backlogs
- Directory control by updating the routing tables of the switching elements of the network based on subscriber movements and requirements
- COMSEC resource control by issuing appropriate directives

The data base also provides the foundation needed to generate reports required by the CSPE and to generate service/agency directives, installation orders, etc. In the initial deployment, the CNCE will be used to aid the CSCE staff in performing its mission until the CSCE is available.

4.10.3.6 Communications Nodal Control Element (CNCE)

The CNCE (AN/TSQ-111) provides an automated capability to exercise technical management and control over nodal communications facilities, including circuit and message switches, trunk transmission systems, network extension facilities, and nodal loop distribution assets. The CNCE is designed to operate as the central facility connecting the interfaces among transmission and switching subsystems and dedicated user systems. The CNCE is more than just another technical control. First, it provides automated techniques for continuous assessment of all nodal equipment. This capability permits real-time fault detection and isolation. Second, its automated controller aids permit rapid determination of corrective actions. These broad capabilities are essential to the improvement of communications circuit availability.

4.10.3.7 Communications Equipment Support Element (CESE)

The CESE represents the lowest level of the TCCF management and control hierarchy and will be designed as an integral part of the equipment or assemblage that it supports. The CESE will provide the broad spectrum of TCCF capabilities required to support equipment operations. Generally, the CESE functions employed in equipment assemblages will provide a capability for the autonomous management, operation, and monitoring of all equipment that may constitute a part of the assemblage (i.e., combiners, multiplexers, cable driver modems, modulators/demodulators, engineering orderwire facilities, COMSEC, radio transmitters, and receivers, BITE, patching facilities, power supplies, etc.). All CESE functions will be performed either automatically or by the operator personnel, utilizing equipment that is an inherent part of the assemblage involved.

COMSEC/Crypto Equipment

The next generation of COMSEC/crypto equipment to be fielded for Air Force tactical ground-to-ground use will consist primarily of members of a family of compatible equipment. All members of that family will share a common, basic subsystem that may be used to provide secure voice. In general, the TRI-TAC COMSEC equipment encrypts secure calls on an end-to-end basis. All trunk groups are bulk-encrypted. Other COMSEC devices compatible with members of the TRI-TAC COMSEC family include devices that will be used for aircraft communications and combat net radio.

4.10.3.8   Sensor/Weapon Data Link Development

The Air Force is currently conducting a number of intensive exploratory/advanced development efforts aimed at achieving major improvements in the data links associated with airborne sensors, airborne weapons, and the stations (airborne and ground-based) that control them. Even though a specific acquisition program is not currently underway to implement the results of these efforts in specific platforms, they are included in this section because of the importance of these communications links in Air Force tactical operations.

The essence of the communications challenge is to achieve sufficiently wide bandwidth to permit system operation in the face of heavy enemy jamming, at a cost that will be low enough to allow the data-link equipment to be installed in thousands of platforms [9]. An exploratory development program is focused on the technology for a multifunction, jamming-resistant, low-cost family of data links. The family is to satisfy a multiplicity of user needs in order to achieve the DoD objectives for standardization and reduction of the proliferation of special-purpose data links. Some user needs include variable data rates, multiple access, ranging, and multipath discrimination [9].

The basic concept for this development program is illustrated in Figure 4.8. Under this concept, multiple, narrow, accurately positioned beams are generated and rapidly switched. The approach should result in sufficient range accuracy and angle accuracy for use on the airborne control platform shown in Figure 4.8. Antijamming and other considerations lead to the use of a combination of frequency hopping and pseudonoise modulation over a wide frequency band. Critical RF problems associated with this approach are being investigated.

Another thrust of the development effort involves designing programmable modems for multiple missions and investigating time-division and code-division multiple access techniques, microprocessors for system control, adaptive arrays, compressed video, and advanced modulation techniques. The overall objective is to control 25 airborne weapons simultaneously from a single airborne platform.

A significant effort is devoted to the signal-processing problems associated with a wideband (> 200 MHz) terminal, which must be mass produced to achieve an affordable unit cost, for example, $55,000. The data rate for the terminal is to be variable, depending upon the amount of jamming protection required and whether or not error-correction coding is utilized.

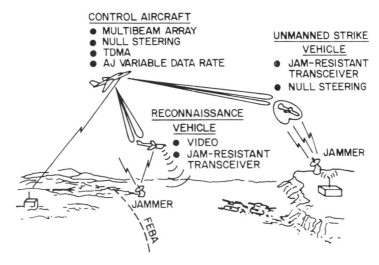

Figure 4.8 Advanced sensor/weapon data link.

4.10.4 Remaining Problem Areas

The planned programs that have been discussed to this point will greatly enhance the capabilities of Air Force tactical communications systems, especially in the areas of security, jam resistance, connectivity, and capacity. Even after these programs have been implemented, however, some important problem areas will remain, especially in the ability of the Air Force's communications equipment to resist a determined electronic warfare and physical attack.

The principal reasons that problem areas will still exist after the current programs have been implemented are (1) the increasing intention of battlefield commanders to attack the communications links of their adversaries; (2) an expected increase in the capability of electronic warfare equipment and antiradiation missiles in the 1990s; (3) changing information-handling requirements associated with new command and control systems; and (4) increased emphasis on providing a viable, survivable, mobile, overseas C^3 capability.

4.11 REFERENCES

1. Thompson, T.H. "Tactical Air Forces Integrated Information System Master Plan." *Signal* (August 1978): 68–73.
2. Eisenberg, R.L. "JTIDS System Overview." Principles Operational Aspects Precision Position Determination System, NATO Advisory Group Aerospace Research and Development. *AGARDograph* 245, (July 1979): 26-1–26-7.
3. Ulsamer, E. "The Growing, Changing Role of C^3I." *Air Force Magazine* (July 1979): 36–48.
4. "TRI-TAC." Joint Tactical Communications Office (TRI-TAC Office), Tinton Falls, New Jersey, September 1979.

5. Conner, W.J. "The AN/TRC-170–A New Digital Troposcatter Communication System." In *Proceedings of New Techniques Seminar, Digital Microwave Transmission Systems*, Department of Electrical Engineering, Princeton University, Princeton, N.J., pp. 19–23, 27 February 1979.

6. Tyree et al., B.E. "Ground Mobile Forces Tactical Satellite SHF Ground Terminals." In *Proceedings of New Techniques Seminar, Digital Microwave Transmission Systems*, Department of Electrical Engineering, Princeton University, Princeton, N.J., pp. 49–53, 27 February 1979.

7. Gleason, M.R. and LaBanca, D.L. "A System Overview of Tactical Satellite Communications." In *Proceedings of New Techniques Seminar, Digital Microwave Transmission Systems*, Department of Electrical Engineering, Princeton University, Princeton, N.J., pp. 30–48, 27 February 1979.

8. Ellington, T. "DSCS III—Becoming an Operational System," pp. 1499–1504.

9. Bush, H. "Multifunction Communications and Tactical Data Links." *Techniques for Data Handling in Tactical Systems II* (AGARD Conference Proceedings 251), Monterey, Calif., October 1978.

4.11.1 Further References

1. Black, K.M. and Lindstrom, A.G. "TACAMO: A Manned Communication Relay Link to the Strategic Forces." *Signal* (September 1978).

2. Blackman, J.A. "Switched Communications for the Department of Defense." *IEEE Transactions on Communications* (July 1979): 1131–1137.

3. Boyes, J.L. "A Navy Satellite Communications System." *Signal* (March 1976).

4. Braverman, D.U. and Waylan, C.J. "LEASAT Communication Services." In *Conference Record of the 1979 International Conference on Communications*, IEEE Pub. 79CH1435-7CSB, p. 33.2.1.

5. Brick, D.B. "Air Force Tactical C³I Systems." In *Proceedings of the 42nd Military Operations Research Symposium*, 6 December 1978.

6. Brick, D.B. and Ellersick, F.W. "Future Air Force Tactical Communications." *IEEE Transactions on Communications* (Special Issue on Military Communications), (September 1980): 1551–1572.

7. Brick, D.B. and Hines, J.W. "The Impact of VHSIC on Air Force Signal Processing." In *Proceedings of the 13th Annual Asilomar Conference on Circuits, Systems, and Computers*, Pacific Grove, Calif., 5 November 1979.

8. Burrows, M.L. "ELF Communications in the Ocean." In *Conference Record of the 1977 International Conference on Communications*, IEEE Pub. 77CH1209-6CSCB, 12 June 1977.

9. Cook, C.E. "Optimum Deployment of Communications Relays in an Interference Environment." *IEEE Transactions on Communications* (Special Issue on Military Communications), (September 1980): 1608–1615.

10. Coviello, G.J. "Comparative Discussion of Circuit-vs. Packet-Switched Voice." *IEEE Transactions on Communications* (August 1979): 1153–1160.

11. Cummings, W.C. et al. "Fundamental Performance Characteristics That Influence EHF MILSATCOM Systems." *IEEE Transactions on Communications* (October 1979): 1423–1435.

12. Davis, J.R., Hobbis, C.E. and Royce, R.K. "A New Wide-Band System Architecture for Mobile High Frequency Communication Networks." *IEEE Transactions on Communications* (Special Issue on Military Communications), (September 1980): 1580–1590.

13. Diffie, W. and Hellman, M.E. "New Directions in Cryptography." *IEEE Transactions on Information Theory* (November 1976): 644–654.

14. Edell, J.D. "Wideband, Non-Coherent, Frequency Hopped Waveforms and Their Hybrids." Naval Research Laboratory Report 8025, 8 November 1976.

15. *Electronics*. Annual Technology Update Issue. New York: McGraw-Hill (25, October 1979).

16. Ellington, T. "DSCS III Planning." *IEEE Transactions on Communications* (Special Issue on Military Communications), (September 1980): 1499–1504.

17. Flanagan, J.L. et al. "Speech Coding." *IEEE Transactions on Communications* (April 1979): 710–737.

18. Fossum, R.R. and Cerf, V.G. "Communications Challenges for the 80's." *Signal* (October 1979): 17–24.

19. Gallowa, R. "U.S. and Canada: Initial Results Reported." In "Optical Systems: A Review," *IEEE Spectrum* (October 1979): 70–74.

20. Gutleber, F.S. and Diedrichsen, L. "TRI-TAC Considerations of Electronic Warfare." *Signal* (March 1978): 52–58.

21. Hamsher, D.H. (Ed.). *Communication System Engineering Handbook*. New York: McGraw-Hill, 1967.

22. *IEEE Communications Society Magazine* (Special Issue on Communications Privacy), November 1978.

23. *IEEE Proceedings on Military Communications*, MILCOM, 1983, Boston, Mass.

24. *IEEE Transactions on Antennas and Propagation*, (Special Issue on Conformal Arrays), January 1974.

25. *IEEE Transactions on Antennas and Propagation* (Special Issue on Adaptive Antennas), September 1976.

26. *IEEE Transactions on Communications* (Special Issue on Satellite Communications), October 1979.

27. *IEEE Transactions on Electronic Devices* (Special Issue on Very Large-Scale Integration), April 1979.

28. *IEEE Transactions on Electronic Devices* (Special Issue on Displays and LED's), August 1979.

29. Kahn, R.E. et al. "Advances in Packet Radio Technology." *Proceedings of the IEEE* (November 1978): 1468–1496.

30. Koga, K. et al. "On-Board Regenerative Repeater Applied to Digital Satellite Communications." *Proceedings of the IEEE* (March 1977): 401–410.

31. Melancon, P.S. and Smith, R.D. "Fleet Satellite Communications (FLTSATCOM) Program." In *Conference Record of the American Institute of Aeronautics and Astronautics*, 8th Communications Satellite Systems Conference, pp. 516, 20 April 1980.

32. Metzger, L.S. "On-Board Satellite Signal Processing." Lincoln Laboratories Tech. Note 1978-2, Vol. 1, 31 January 1978. This report is available from the National Technical Information Service, Doc AD-AO52768.

33. Munson, R.E. "Conformal Microstrip Antennas and Microstrip Phased Arrays." *IEEE Transactions on Antennas and Propagation*, (January 1974): 74–78.

34. Nooney, J.A. "UHF Demand Assigned Multiple Access (UHF DAMA) System for Tactical Satellite Communications." In *Conference Record of the 1977 Conference on Communications*, IEEE Pub. 77CH1209-6, pp. 45.5-200, 12 June 1977.

35. Oetting, J. "An Analysis of Meteor Burst Communication for Military Applications." *IEEE Transactions on Communications* (Special Issue on Military Communications), (September 1980): 1591–1601.

36. *Proceedings of the IEEE* (Special Issue on Adaptive Arrays), February 1976.

37. *Proceedings of the IEEE* (Special Issue on Microprocessor Applications), February 1978.

38. *Proceedings of the IEEE* (Special Issue on Fault-Tolerant Digital Systems), October 1978.

39. *Proceedings of the IEEE* (Special Issue on Packet Communications Networks), November 1978.

40. *Proceedings of the 6th Annual International Conference on Fault-Tolerant Computing*, Pittsburgh, Pa., 1976.

41. Ricardi, L.J. "Communication Satellite Antennas." *Proceedings of the IEEE* (March 1977): 356–369.

42. Roberts, L.G. "The Evolution of Packet Switching." *Proceedings of the IEEE* (Special Issue on Packet Communications Networks), (November 1978): 1307–1313.

43. Rosen, P. "Military Satellite Communications Systems: Directions for Improvement." *Signal* (November/December 1979): 33–38.

44. Schoppe, W.J. "The Navy's Use of Digital Radio." *IEEE Transactions on Communications* COM-27 (December 1979).

45. Stein, K.J. "Forward-Looking Technology Explored." *Aviation Week & Space Technology* (29 January 1979): 195–202.

46. Sugarman, R. "Superpower Computers." *IEEE Spectrum* (April 1980): 28.

47. Sunmey, L.W. "VLSI with a Vengeance." *IEEE Spectrum* (April 1980): 24.

48. Sussman, S.M. "A Survivable Network of Ground Relays for Tactical Data Communications." *IEEE Transactions on Communications* (Special Issue on Military Communications), (September 1980): 1616–1624.

49. Swyre, D.G. "The IDCSP-Concept and Performance." In *Conference Record of the American Institute of Aeronautics and Astronautics*, 3rd Communications Satellite Systems Conference, Los Angeles, Ca., AIAA Paper 70-492, April 1970.

50. Tyree, B.E. et al. "Ground Mobile Forces Tactical Satellite SHF Ground Terminals." In *Proceedings of New Techniques Seminar, Digital Microwave Transmission Systems*, Department of Electrical Engineering, Princeton University, Princeton, N.J., pp. 49–53, 27 February 1979.

51. van der Meulen, E.C. "A Survey of Multi-Way Channels in Information Theory: 1961-1976." *IEEE Transactions on Information Theory"* (January 1977): 1–37.

52. Wayland, G.J. and Yowell, G.M. "Considerations for Future Navy Satellite Communications." In *Conference Record of the Electronics and Aerospace Systems Conference*, IEEE Pub. 79GH1476-1 AES, p. 623, 11 October 1979.

53. Weiner, T.F. and Karp, S. "The Role of Blue/Green Laser Systems in Strategic Submarine Communications." *IEEE Transactions on Communications* (Special Issue on Military Communications), (September 1980): 1602–1607.

54. Weinrich, A.W. "The DSCS III Communication Satellite Performance." In *Conference Record of the 1977 International Conference on Communications*, p. 291, June 1977.

55. White, T.F. ''Fleet Satellite Communications/Leased Satellite Communications Operations.'' In *Conference Record of the 1979 Conference on Communications*, IEEE Pub. 79CH1435-7CSB, pp. 33.3.1, 10 June 1979.

56. Wright, I.C. and McLellan, P. ''The Defense SATCOM System, Phase II—An Operational System.'' In *Conference Record of the 1977 International Conference on Communications*, pp. 31.1-281, June 1977.

Chapter 5

SATELLITE COMMUNICATIONS/ CONTROL

Satellite communications are used throughout all phases of military activity—from peacetime, through minor conflicts, conventional war, and nuclear war. Both civilian domestic and military-operated satellite systems provide for critical voice, record, teletypewriter, video, and data communications for long-haul connectivity.

Numerous books have been devoted to the subject of satellite communications [1]. This chapter will present a summary of relevant commercial and military satellites as well as a discussion of satellite control techniques including a description of the new Consolidated Space Operations Center (CSOC). Since considerable interest is being shown for the utilization of commercial satellite systems by the military, some discussion of their use and characterization is also provided.

5.1   MILSATCOM

In the future, there will be considerably more reliance on satellite systems for long-haul narrow and wideband secure communications. The DoD MILSATCOM program provides for a comprehensive satellite architecture supporting strategic/tactical connectivity to the forces on land, sea, and in the air. These systems will provide antijamming (AJ) and low-probability-of-intercept (LPI) capability as never before envisioned.

The key elements of the MILSATCOM architecture are:

- MILSTAR
- AFSATCOM
- DSCS II/III
- FLTSATCOM
- Defense Meteorological Satellite Program (DMSP)
- Satellite Defense System (SDS)

- Global Positioning System (GPS)
- Defense Satellite Program (DSP)

These systems will provide long-haul connectivity for the military as well as weather data, navigational information, and surveillance data. When the MIL-STAR system becomes operational in the late 1980s, it will provide critical voice/data capability in nuclear and postnuclear warfare situations.

The development of satellite systems is such that MILSTAR typifies the evolution to higher and higher frequencies (from UHF through SHF to EHF) for wider bandwidth and LPI. In addition, there is continued application for AJ capability utilizing spread-spectrum modulation such as frequency hopping (FH). The next generation of military satellite systems will be much more robust than present systems.

The evolution of the MILSATCOM systems through the 1990s is illustrated in Figure 5.1. The wideband user community is comprised of a diverse mix of fixed, transportable, shipboard, and airborne terminals. It is served by the DSCS system at SHF, presently using the DSCS-II satellites with planned transition to satellites that will provide significant improvements in jam resistance, capacity, flexibility, in-orbit life, and system control. The DSCS III upgrade may include EHF transponders for providing enhanced jam resistance and high data-rate capabilities to selected wideband users.

The tactical mobile community consists of a large population of mobile platforms operating with small antenna terminals. Initially served at UHF by the

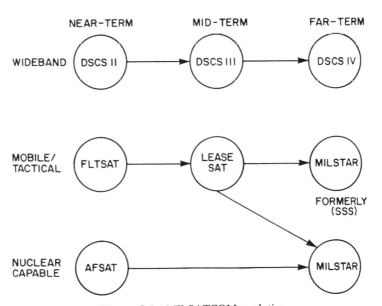

Figure 5.1 MILSATCOM evolution.

GAPFILLER and FLTSAT spacecrafts, these satellites are being gradually replaced by leased satellites (LEASATs). However, a UHF space-segment capability will probably be required well into the 1990s in view of the heavy DoD investment in UHF terminals. A new system, TACSATCOM II, is proposed for the far term (1990s) to alleviate the deficiencies of UHF systems in the areas of bandwidth and jam resistance. The TACSATCOM II will operate primarily in the EHF band, with on-board processing and, most likely, antenna nulling for improved jam resistance.

The nuclear-capable forces community is presently served by the AFSATCOM system consisting of communication packages on a number of satellites in geostationary and elliptical orbits. In the near term, the present AFSATCOM system will be supplemented by the addition of single-channel transponders (SCTs) on DSCS III and other host satellites. The Strategic Satellite System (SSS), proposed for the mid-1980s, included an evolving capability in the far term. A transition to the SSS system with enhanced jam resistance and physical survivability is considered essential for the vital users of the system.

The SSS, which is now in development as MILSTAR, is designed to provide critical command and control communications from the national command authorities to the CINCS of the unified and specified commands and to the nuclear-capable forces. It will be designed to survive and endure through all phases of a protracted nuclear engagement as one component of a battle management system for the command and control of strategic forces.

The progressive improvement proposed for each of the MILSATCOM systems requires a vigorous research and development program to define, develop, and test advanced concepts and technology, preferably in a quasi-operational environment.

5.1.1   Wideband Users/Systems

Enhanced antijam protection for the high data-rate users is of major importance in the far-term evolution of the DSCS system. To accomplish this objective, it was recommended that the high data-rate users be moved into the higher 30/20 band. A gradual time-phased transition into the 30/20 GHz band can be achieved initially through the addition of straight-through transponders on DSCS III satellites in the late 1980s and more complex processing transponders on later satellite designs (DSCS IV). The 8/7 GHz band, however, will continue to remain the primary band for DSCS operation even in the DSCS IV era.

An interservice coordinated R & D program in the 30/20 GHz band is necessary for the development of system operation concepts and the space and ground segment technologies. With regard to the satellite multiple-access technique, TDMA appears to be the preferred approach with the DSCS III satellites, since it would provide a higher throughput than FDMA while avoiding the transponder intermodulation and uplink power control problems associated with FDMA. More advanced multiple-access techniques, such as satellite-switched TDMA (SS-

TDMA), TDMA with time-hopped beams, etc., are candidates for application with the DSCS IV satellites. Under benign conditions, an efficient multiple-access scheme can increase the satellite's data handling capacity substantially and provide more flexibility for accommodating users with differing data rates and connectivity needs.

The DSCS III satellite will use narrow-beam antennas for 30/20 GHz operation. For better jam resistance, the feasibility of a 30 GHz receive nulling antenna, such as a multibeam antenna (MBA) or a phased array on DSCS III, should be examined. In the long term (DSCS IV), antenna discrimination from the satellite will be essential for high data-throughput under strong uplink jamming. Antenna nulling techniques and their hardware implementation will be investigated for operation with SS-TDMA and time-hopped beams for potential DSCS IV application.

In the 30/20 GHz band, 1.5 GHz of bandwidth is allocated internationally for fixed and mobile satellite service. In the United States, however, the upper 1 GHz of the exclusive band is assigned to the DoD. Since it would be difficult to spread pseudonoise waveforms over 1 GHz, conventional frequency-hopping, frequency-hopped TDMA, and hybrid (frequency hopping in combination with pseudonoise) schemes may be explored for users requiring antijam (AJ) protection. The selected AJ technique will be used in conjunction with antenna nulling for the DSCS III and onboard AJ processing in the far term with DSCS IV satellites.

If the decision is made that the DSCS IV satellite should use SS-TDMA or something similar, it would be necessary to provide the capability of beam switching (appropriate interconnection of uplink and downlink satellite antennas) in the satellite. The switching can be done either at RF or at baseband. Baseband switching, although more complex (since it requires complete demodulation of the uplink signals in the satellite), would provide greater dynamic flexibility for interconnecting users distributed over a wide geographical area.

The need for satellite onboard AJ processing at 30/20 GHz will depend largely on the characteristics of potential user communities and their terminal capabilities. For example, earth terminals with large antennas (e.g., 40 ft paraboloid) can operate effectively even under heavy uplink jamming, with conventional straight-through transponders. A processing transponder will improve their AJ performance by only 6 dB, which probably would not justify the additional hardware processor complexity on the spacecraft. On the other hand, for small-antenna mobile/transportable terminals, onboard-processing will be essential for supporting even 2400 bits/s under strong uplink jamming. The issue of onboard-AJ processing will be deferred until a determination of 30/20 GHz users and their terminal characteristics is established.

Research and development efforts will need to focus on the necessary technology for the 30/20 GHz spacecraft and user terminal hardware components. Primary emphasis will be placed, in the near term, on the development of DSCS III-U spacecraft components and associated user terminals.

5.1.2   Tactical Mobile Users/Systems

The next-generation tactical satellite communications system, TACSATCOM II, will operate in the EHF frequency band with satellite onboard processing and (perhaps) antenna nulling for maximum jam resistance. It will also have LPI. The MILSATCOM Architectural Working Group has recommended the use of the 45 GHz band for the uplink; the associated downlink still remains to be determined. While the 40, 20, and 7 GHz bands are all candidates for the downlink, the 20 GHz appears to be the best compromise, on the basis of simpler frequency/orbit coordination than 7 GHz, less rain attenuation than at 40 GHz, and adequate jam resistance against downlink jamming.

For most tactical users, introduction of TACSATCOM II will require a transition from present operation at UHF to operation at EHF. Reequipping tactical platforms with new terminals is both a long-term and expensive proposition. In many cases, users may have to operate simultaneously at both UHF and EHF during the transition phase. Timing of the transition from the LEASAT to TACSATCOM II will depend largely on the availability of EHF terminals. In planning for this transition it must be determined whether the TACSATCOM II payload design must also support the entire existing UHF communication requirements prior to UHF/EHF transition or whether LEASAT-like assets would be maintained for separate UHF communications beyond the presently programmed schedule.

5.1.3   Nuclear-Capable Users/Systems

The general characteristics and needs of the nuclear-capable users community are similar to those of the tactical/mobile users. The nuclear-capable users, however, require maximum physical and electronic survivability under extreme conditions for command and control of forces. The Strategic Satellite System (SSS-MILSTAR) serving this community will evolve from its present capability (AFSATCOM) through a more jam-resistant configuration with better physical survivability to a system with substantial jam resistance and physical survivability in the 1990s.

One possible candidate for the MILSTAR system is a set of dedicated satellites in inclined circular orbits at five-times synchronous altitude, together with single-channel transponders (SCTs) on a number of host satellites in both geostationary and elliptical orbits. Because of sunk cost in UHF force terminals, the uplinks from small mobile platforms probably will continue to be at UHF. Selected users may be provided with SHF and/or EHF uplinks. The downlink from the satellite will be at UHF and SHF. Another candidate being studied includes a number of proliferated satellites.

Significant technology development effort for the SSS does not appear to be necessary. Technology maturity has been achieved through the comprehensive Research Development Test and Evaluation (RDT & E) program, leading to the operational test and evaluation of the Leased Experimental Satellite (LES) 8/9 satellites with prototypical R & D terminals. These demonstration tests validated critical design concepts such as crosslinks, advanced jam-resistance modulation techniques, and the use of EHF frequencies. A sound technological base is extant to support the evolution of the SSS system through the 1980s.

5.2   CONSOLIDATED SPACE OPERATIONS

Recently, the Air Force and other services have recognized the need for more communications in space and have created a space command where the dominant mode of communications will be via satellites, domestic and DoD, through the space shuttle and other NASA resources. The Consolidated Space Operations Center (CSOC) will be a major element in the control and resource allocation for the combined DoD/NASA satellite system.

For military systems, survivability, improved flexibility, and the need for service to small mobile terminals are the principal factors involved. Technical trends include the use of higher frequency bands, multibeam antennas, and a significant increase in the application of onboard processing. Military systems will employ a variety of techniques to counter both physical and electronic threats. The use of redundant transmission paths is a particularly effective approach. Successful implementation requires transmission standards to achieve the required interoperability among the pertinent networks. For both the military and commercial sectors, the trend toward larger numbers of terminals and more complex spacecraft is still persisting [2].

The drive for the development of future commercial systems is influenced by the pervasion of digital techniques and services, growing orbit and frequency congestion, the demand for more entertainment, and the large potential market for commercial "roof-top" service.

Since the 1960s, commercial and military applications of satellite communication have continued to exploit the unique attributes of the medium. These include: distance-insensitive cost, higher quality transmission and higher data rates for long-range systems or mobile service, and ease of establishing multipoint network connectivity. Early system configurations involved simple, lightweight spacecraft and a small number of large fixed terminals providing global multichannel trunking. As new applications have emerged, satellite complexity and the terminal population have increased dramatically.

5.3   CURRENT STATUS

Table 5.1 shows the frequency bands allocated for satellite communication and Table 5.2 summarizes the relevant characteristics of the current space and earth segments used by the military and commercial sectors. Virtually all the satellites are in geosynchronous orbit. The discussion of domestic carriers is limited to

Table 5.1 Frequency bands generally employed for satellite communication.

Popular Designation	Range in MHz	Type Service
Military UHF (In U.S. and Possessions)	240.0–328.6 Up and Down 335.4–399.4 Up and Down	Mobile-Satellite* (LES-5,6:TACSAT I:GAPSAT:FLEETSAT)
L	1,535.0–1,543.5 Down 1,636.5–1,645.0 Up 1,542.5–1,558.5 Down 1,644.0–1,660.0 Up	Maritime Mobile Satellite (MARISAT) Aeronautical Mobile-Satellite
C	3,700.0–4,200.0 Down 5,925.0–6,425.0 Up	Fixed-Satellite* (INTELSATS, COMSATS, REGIONAL)
X or Military SHF	7,250.0–7,775.0 Down 7,900.0–8,400.0 Up	Fixed-Mobile-Satellite* (DSCS, NATO)
K_u	10,950.0–11,200.0 Down 11,450.0–11,700.0 Down 14,000.0–14,500.0 Up	Fixed-Satellite* (INTELSAT V, SBS)
K_a	17,700.0–21,200.0 Down 27,500.0–31,000.0 Up	Fixed-Satellite* (Japanese CS)
Q	40,000.0–41,000.0 Down 43,500.0–47,000.0 Up	Fixed-Mobile-Satellite
V	50,000.0–51,000.0	Fixed-Satellite

*Shared with terrestrial communication.

Table 5.2 Current satellite communication assets.

| System (military) | Freq. Band (GHz) | Satellites | | | | Terminals | |
		Number and Designation	Wt.(#)	Power (W)	Approx. Number	Transmit Power (W)	Antennas
Wideband	"X" or "Military SHF" (7–8)	(7) DSCS II (1) DSCS III	1350 2520	520 1100	250	500–10,000	6 ft to 60 ft
Tactical Mobile	"Military UHF" (0.24–0.40)	(4) FLEETSAT (3) GAPSAT Leased Transponders	2300	1650	1000	100	Mostly shipborne, 4-element arrays
Strategic Mobile	"Military UHF" (0.24–0.40)	AFSAT payloads on host spacecraft	—	—	500	100	Mostly airborn, simple structures
(Commercial) INTELSAT	"C" Band (4–6) "K$_u$" Band (11–14)	(7) IS IV-A (5) IS V	1820 2130	590 1475	400	10,000	30 ft–110 ft
Domestic Carriers	"C" Band and "K$_u$" Band	(17) COMMSTAR (ATT/GTE) WESTAR (WU) SATCOM (RCA) Sat Business Syst ANIK (Canada) A B C	1790 655 1010 1220 655 1016 1210	760 305 745 900 330 840 1000	600 (2-way) >10,000 (TV rec. only)	500–10,000	15 ft–100 ft
Maritime Service	"L" Band (1.5–1.6)	(3) MARISAT INMARSAT transponder on IS-V Flt 5	720 —	355 —	1500	50	Shipborne 4 ft

the western hemisphere. In the next few years, five additional carriers expect to have satellites in orbit, raising the total number of domestic satellites to about 30.

Multichannel voice trunking still dominates the commercial systems, with an annual growth rate of 20 percent for INTELSAT. TV distribution for cable networks, education, and health services is growing in importance for domestic systems. Other current applications include radio voice distribution, business data networks, and leasing of point-to-point voice and data circuits with rates ranging from 2400 BPS to about 10 Mbps. Only about 3 percent of the trunks in the AT&T switched network are carried by satellite because of (1) objections to the time delay caused by more than a single hop, (2) the opportunity to expand capacity at competitive costs by overbuilding on existing microwave relay and cable paths, and (3) the potential of fiber optics transmission. Competing carriers offering long distance telephone service are finding satellite systems cost effective for expanding their trunk coverage to new geographic areas. Video and audio conferencing as a cost-effective substitute for business travel is offered by several carriers. Current military applications include a global fixed service at X-band for voice and message traffic and high data-rate transmission and mobile service using the military UHF band for tactical and strategic command and control.

The commercial satellite payloads using C-band are generally arranged as a group of 12 or 24 transponders (depending on whether or not dual polarization is employed), with transponder bandwidths ranging from 35 to 40 Mhz. This permits about 2,000 two-way voice channels per transponder using FDM FM, the prevailing multiplexing and modulation technique. However, recent development with FDM-SSB-AM indicates that this capacity can be doubled. Other modulation techniques in current use include single channel per carrier (SCPC) for low traffic density routes and DPSK for digital transmission service. The multiple access technique is most generally employed in FDMA, but TDMA has been introduced using burst rates from 1.5 to about 60 Mbps. Commercial satellite antennas provide earth coverage beams or beams approximately shaped to serve specific land masses. Some fixed spot beams are employed by INTEL-SAT.

The military X-band satellites are configured with a small number of transponders (four on DSCS II and six on DSCS III). They serve a variety of users with a wide disparity of terminal sizes, data rates, and modulation techniques. FDMA is employed. Antenna coverage is very flexible, providing earth coverage and movable spot beams. DSCS III also has a capability for antenna nulling to provide antijamming (AJ) protection, and multibeam uplink and downlink coverage, to concentrate power where required. The military UHF service (e.g., FLEETSAT) provides ten transponders with a 25 KHz bandwidth, one transponder with a 500 KHz bandwidth, and another that multiplexes 12 channels operating at 75 BPS and employing FSK modulation. The 25 KHz channels generally carry rates from 2400 to 9600 BPS using DPSK modulation. The military UHF antenna beams provide earth coverage. Spread-spectrum transmission is employed in both the X and military UHF bands. Both direct sequence (pseudonoise) and frequency hopping techniques are used. The UHF service includes onboard processing.

5.4 FUTURE SYSTEMS

The single most important influence on future commercial systems is the explosive pervasion of digital techniques and services in both satellite and terrestrial systems. With the advent of the "Information Age" encompassing electronic mail, word processors, paperless offices, and electronic funds transfer, a $20 billion per year data communications market is forecast, with about four million business computers expected by 1990.

Digital integrated circuit technology, and the low-cost electronic devices that it creates, are beginning to revolutionize communication networks. A new "synergy" among the switching, transmission, and process segments permits a vastly improved efficiency of channel utilization and significant reductions on investment, operation, and maintenance costs. An example is the AT&T ESS-4 electronic switching central, which processes over 100,000 telephone trunks. The use of digital modulation for transmission is often less costly than other alternatives and is better suited for many channel and device characteristics commonly encountered in both satellite and terrestrial circuits, such as nonlinearity, intermodulation products, and interchannel crosstalk. Digital techniques are also well adapted for built-in testing, flexible channel multiplexing, and such complex signal manipulations as echo-cancellation or time-assigned speech interpolation (TASI).

The use of digital transmission and techniques is also increasing rapidly in military systems. The need for secure voice and record traffic for both fixed and mobile service is an important contributor to the trend. Also, a number of processes used to protect against jamming, interception, or a fading transmission medium (including frequency hopping, forward error correction coding, interleaving, etc.) require digital processing procedures.

A second important influence on future commercial systems is the congestion in synchronous orbit and its limiting effect on total transmission throughput. Nominal satellite spacing is 3° in longitude, based on the ability to discriminate with medium-sized terminals (e.g., 15 ft at C-band). Locations in the useful portion of the geostationary arc are saturated at C-band and rapidly filling at K_u band. The tendency toward the use of smaller terminals, and hence wider antenna beamwidths, will exacerbate the situation. Alternatives to alleviate the frequency-orbit congestion include the utilization of higher frequencies (e.g., K_a band), frequency reuse by spacecraft with many narrow antenna beam positions, increased channel utilization efficiency by dynamic allocation processes, and large platforms that concentrate an aggregation service capability in one location. The trend toward higher frequencies is slow because of concerns about the need for development of new devices, the effects of rain on attenuation and depolarization, and the higher cost of terminals.

There is a need for rain margin as carrier frequency varies and for frequency sensitivity to other factors such as look angle, required channel availability, and amount of local rainfall. Another approach for dealing with outages due to rain attenuation is the use of diversity terminals separated by approximately 10 miles.

Several new commercial satellite markets are receiving careful attention. The public's insatiable demand for more entertainment has generated a flurry of activity aimed at direct TV broadcast to potentially millions of home receivers costing only a few hundred dollars. The Direct Broadcast Satellite (DBS) Service envisions the use of the 11.7–12.6 GHz band. Typical system design approaches consider a satellite transmitter power of approximately 200 watts in conjunction with antenna beams shaped to approximate each time zone coverage pattern. The home receivers are expected to utilize an antenna diameter of about 2.5 ft and employ a front end with a 4.5 dB noise figure. Proposed service includes standard color video and high fidelity audio.

Perhaps the largest potential market is customer-premises service for tens of thousands of commercial roof-top terminals, competing with terrestrial "tail-circuits." User requirements vary, but the average may be typically a few voice or 64 Kbps data circuits. A smaller population of high-density data users could employ a 1.5 Mbps. Many studies of this type of application have been conducted under NASA sponsorship. The concepts generally considered large spacecraft (e.g., 5,000–10,000 lbs), multibeam antennas involving up to hundreds of beam positions using beam sizes in the range of 0.3 to 0.5°, and terminals with a few tens of watts and a 10 to 15 ft diameter antenna. Terminal cost is the pivotal issue. It should be well below $100,000 to permit the service to compete with terrestrial systems.

Encryption techniques are expected to be incorporated in future commercial systems for several reasons. Computer data encryption will be used to prevent unauthorized use of commercial material, and some form of scrambling will be employed to discourage unauthorized reception of entertainment broadcasts. It is also likely that future satellite control systems will incorporate improved security.

Future military systems are influenced by many needs not generally considered for commercial networks. Foremost among them is survivability in the face of any possible enemy threat. The threats include uplink or downlink jamming, position location for tactical exploitation, interception of messages, "spoofing" of the communication channels or the satellite control system, physical destruction of the earth or satellite nodes, and several types of effects resulting from detonation of nuclear weapons. The latter includes destruction or degradation in electrical performance of the components and devices in both the space and terrestrial segments. Electromagnetic pulse (EMP) is singled out as a particularly threatening hazard for terrestrial nodes. Other concerns about nuclear weapon effects include the fading and scintillation produced by transmission through a disturbed medium such as the layers of the ionosphere.

Another impetus unique to military systems is the need for wide flexibility in antenna coverage, capacity allocation, network coverage and configuration, and the capability for rapid extension of service to new areas. Since a large majority of the military terminals are mobile, and therefore disadvantaged in terms of power and antenna gain, the burden tends to fall on the satellite configuration to provide the required performance characteristics (i.e., high ERP and G/T). This requirement tends to make military satellites more complex and more expensive than commercial satellites.

Leased commercial satellite service is employed by DoD in a number of applications. Those applications include overseas and domestic telephone trunks and a number of wideband circuits on domestic systems. The current U.S. policy regarding the ability to conduct a protracted war, and the need for preservation of continuity of government, requires a resilient communication system that can survive all levels of conflict and be capable of reconstitution. Because of the large government dependence on commercial systems in the continental United States, an advisory committee has been appointed by the President to investigate the possibility of funding commercial systems to upgrade their survivability where appropriate.

5.5 TECHNICAL TRENDS AND ISSUES

The evolution to higher frequency bands is usually paced by the development of new components and devices such as RF power amplifiers and low noise receivers. The technology is maturing rapidly up through the K_u band for commercial applications. Also, solid state power amplifiers in the 5 to 8 watt range are beginning to replace traveling wave tubes at C-band. NASA is currently conducting a broad-based R & D program focusing on K_a band utilization, and is planning to fly an advanced communication technology satellite employing high-gain multibeam antennas with onboard switching. For future military mobile service, EHF (44 GHz) up and SHF (20 GHz) down will be employed in order to obtain improved antijamming protection, reduced scintillation and fading in a nuclear-disturbed medium, and relief from the spectrum congestion experienced at lower frequencies. Satellite-to-satellite crosslinks will be introduced to minimize dependence on ground relay stations for very long distance links.

Increased use of large multibeam antennas is expected on future satellites for both commercial and military applications. Approaches involving reflectors, as opposed to lenses, are generally preferred, so far. A major technical concern is the control of the sidelobe levels, especially for scan angles relatively far (~ 10 beamwidths) from the boresight. Other important issues to be resolved include crowding of the feed plane and control of the feed patterns. The use of a few scanning beams, as opposed to many simultaneous beams, would ease the sidelobe requirements and offer a good solution for providing service to areas with a very nonuniform geographic traffic distribution. Since scanning beams can be generated with antenna arrays, they offer more flexibility in design approaches; however, they require the use of TDMA, which presents both advantages and disadvantages.

For service to identical terminals with uniform geographic distribution and receiving equal data rates, a satellite with N beams operating with FDMA uses the same total transmitter power as one with a single beam hopping to the N locations and synchronized with TDMA. However, for the uplink, more terminal ERP (by the factor N) is required for TDMA for the same link margin. Other disadvantages of TDMA include higher terminal costs due to the complexity of the modems, increased bit storage at the terminal, and more complex synchronization procedures. There are several offsetting advantages of TDMA. If onboard satellite switching among many beam locations is planned, it offers a vastly simpler design approach for the filter and matrix switch configurations. On the other hand, its application can be used to advantage in accommodating nonlinear device characteristics, distribution of satellite capacity to terminals with vastly different or time-varying traffic requirements, and in demand-assignment arrangements for dynamic allocation of capacity. For both military and commercial service, hybrid combinations of FDMA and TDMA are expected, depending on cost-trades and specific services to be provided.

Significantly increased use of onboard processing can be expected in future satellites. For commercial systems, switching among channels and antenna beams, at either RF or baseband, will be a principal application. Current designs include an 8×8 matrix. For the customer-premises service described above, there may be the need to switch hundreds of ports. Another expected application of onboard processing is adaptive control of individual beam locations serving areas where extra rain margin is required. Military uses of processing will include demodulation, regeneration, and remodulation of spread-spectrum systems to protect small terminals from downlink power robbing by an uplink jammer; error coding and interleaving of data sequences; reconfiguration of antenna beam coverage for nulling out jammers or redistributing patterns; switching time or frequency slot assignment to alter network connectivity as required by changing scenarios; and establishing half-duplex or full-duplex switching connections for ad hoc conferencing.

Because of the wide spectrum of potential enemy threats that military systems must address, a broad-based approach to survivability is necessary. One very effective technique is the use of redundant paths and alternate transmission media for connecting critical network nodes. Implementing this approach requires interoperability among the transmission links of the diverse networks involved. Link characteristics to be standardized include carrier frequency, type of modulation, data rates, baseband signal formats, methods of acquisition and synchronization, etc.

Progress in achieving interoperability among diverse military terrestrial systems has been slow because of institutional issues. The initial development of a new satellite communication system in a new frequency band offers a unique opportunity to pursue standardization. Interoperability is established with commercial carriers where it is cost effective or convenient. Progress is limited by fear of inhibiting innovation and hence competitive advantage. With the heavy government usage of domestic carriers, improved interoperability will be sought as an important means for upgrading survivability. Candidate rates for standardizing include 75, 2400, 9600, 16,000, and 56,000 BPS and 1.544 MBPS (the DS-1 rate). There is a tendency to use the DS-1 rate as a de facto standard among the satellite carriers. For higher rates, there is virtually no interoperability.

The use of packet switching is increasing in terrestrial systems that provide service to customers with "bursty" traffic, e.g., interactive computer and terminal applications such as ticket reservation systems. With the projected increased use of computers in both the military and commercial sectors, proponents of this technique are predicting much greater utilization. Another advantage of packet switching is the ability to integrate systems with diverse transmission characteristics (carrier frequencies, data rates, etc.), thus providing an alternative approach to interoperability. This choice depends, of course, on the agreement by all concerned parties to standardize on the packet configuration. A central issue here is how much this switching technique can do for conventional voice circuits that are only slightly bursty and are expected to continue to dominate the total traffic. Experiments are continuing using both satellites and ground stations as network nodes.

As satellite communication systems continue to proliferate, new applications can be expected to emerge. Some will be created as a result of the unique characteristics of the medium. Others will be shared by terrestrial systems, especially where the relevant advantages and disadvantages tend to be complementary. In many cases, closer integration of satellite and terrestrial systems will occur, particularly in applications involving digital transmission and processing. A basic technical challenge for the foreseeable future will be the search for minimum cost approaches involving very complex spacecraft serving a very large number of user terminals.

5.6 SATELLITE COMMUNICATIONS LINK REQUIREMENTS AND CAPABILITIES

The military satellite systems of the future must provide the capabilities to meet the stringent requirements of both the conventional and nuclear war scenarios. That capability is not easy to accomplish technologically. This section summarizes the capabilities of some of the present and future satellite systems.

Table 5.3 shows some of the capabilities of MILSTAR, the Defense Satellite Communications System (DSCS), the Satellite Defense System (SDS), the Air Force SATCOM (AFSAT), Fleetsat (FLTSAT), the Defense Satellite Program (DSP), and the Global Positioning System (GPS). Downlink/uplink frequency, AJ coverage, and data-rate capabilities are summarized.

The future need for enduring communications connectivity to the tactical and strategic forces encourages the use of satellites as a long-haul means of communication. However, as was pointed out earlier, the Achilles heel is that this form of communication also has unique vulnerabilities to physical destruction, jamming, and interception.

Table 5.3 Satellite communications capability.

Capability	MILSTAR	DSCS (SCT/ECCM)	SDS (SCT/AFSATI)	FLTSAT	DSP MDM SCT	GPS SCT
Downlink Frequency						
SHF	X	X			X	X
UHF	X	X	X	X	X	X
Uplink Frequency					NA	NA
EHF	X					
SHF		X	X			
UHF	X	X	X	X		
AJ Capability		SCT ECCM	SCT AFSATI			
uplink	X	FH PN	FH PN			
downlink	FH	FH PN	FH	FH	FH	FH
Single Sat. Coverage	EC Equatorial plus Elliptical orbit	EC Equatorial orbit	EC Elliptical orbit	EC Equatorial orbit	EC	EC
Crosslinks	X				X	
Worldwide Connectivity	Provided	Requires Relay	Requires Relay	Requires Relay	Provided	Requires Relay
Data Rate Capability*						
Narrow Band	X	X	X	X	X	X
Wideband	X	X		X	X	
Multiple Access	TDM/FDM	TDM	TDM	WB-CDM		

*Narrow band: 75 b/s. Wideband: multiple 16 K b/s or 2.4 K b/s

This section summarizes the capabilities of current and future satellite systems with an emphasis on demand assignment of network channel capacity and antijam protection. In the future, both terrestrial and satellite links will take advantage of the efficiencies gained in time-division multiple access (TDMA) and control techniques that ensure enduring communications connectivity.

As Table 5.3 indicates, only the MILSTAR system will be equipped to handle EHF frequencies. Perhaps in the future the Defense Satellite Communication System (DSCS) will accommodate EHF transponders. The table does indicate a trend toward the higher frequencies, EHF, with less reliance on UHF. However, many of the currently operational satellites (e.g., FLTSAT) are predominantly UHF. Numerous studies have shown that this is not an adequate frequency range for communications in hostile environments, especially nuclear. However, the terminals are inexpensive.

Table 5.3 also indicates that there will be more AJ protection in future satellites using either psuedonoise (PN) or frequency hopping (FH) spread spectrum. Worldwide connectivity can be achieved only through relays in most satellites with the exception of MILSTAR and DSP. Crosslinking via lasers or other links is something new that will be accommodated in the MILSTAR system.

In many of the current systems, only narrowband capability (75 b/s–2.4 Kb/s) is readily available. However, in future systems, multiple 16 Kbps channels will be available for DoD use. Wider bandwidths to 30 Mb/s will also be available in selected satellite systems. Multiple access using time-division multiple access (TDMA) will be available in the MILSTAR system.

5.7 THE MILSTAR SYSTEM

Until recently, only scant attention has been paid to the vulnerability of satellite communications systems because military strategists believed a nuclear war would be over in a matter of hours. Planners now believe that a nuclear war could more likely be protracted. Consequently, strategists have embarked on a crash program to upgrade and harden its present systems; however, even the improved systems will remain vulnerable. That is why development of MILSTAR is being given top priority.

The program's general shape is set, but final details are far from resolved; e.g., the specification of the MILSTAR terminals is not yet final. The success of the system will depend on development of more advanced communications technology.

To ensure global coverage, the MILSTAR system, as presently envisioned, will comprise seven satellites: four in geosynchronous orbit over the equator and three in highly elliptical polar orbits that would ensure continuous coverage for bombers and submarines operating near the North Pole, where the geosynchronous satellites would be below the horizon.

MILSTAR satellites will be the ''smartest'' yet planned. They will be switching centers in space, controlling access to the system and routing messages to end users. They will search for potential users automatically, scanning the earth with narrow antenna beams that hop from spot to spot. They will also be the first satellites that can function autonomously. If the control centers on earth were destroyed, MILSTAR satellites could maintain their position for months.

Besides being able to establish the usual uplinks and downlinks of communications satellites, MILSTAR will be able to create crosslinks among the satellites themselves—a capability recently introduced in civilian communications satellites. The crosslinks will enable MILSTAR satellites to relay messages around the world above the atmosphere without reliance on intermediate downlinks to vulnerable ground stations. This capability would ensure, for instance, that a presidential jet flying over the Pacific would be able to communicate with a tactical nuclear force in Europe even if all ground stations had been destroyed.

MILSTAR crosslinks will operate at even higher frequencies—around 60 GHz. Use of this frequency range will prevent eavesdropping and interference from enemy ground stations, since the atmosphere is nearly opaque to 60-GHz signals.

MILSTAR will also be able to communicate at ultra-high frequencies (UHF), where most conventional military communications systems operate. Broadcasting messages is one application envisioned for this compatibility feature. For example, a flying command post equipped with a MILSTAR terminal would be able to transmit a message up to MILSTAR at EHF to elude jamming. The satellite could then transmit the message back down to earth at both EHF and UHF wavelengths. This capability would enable non-MILSTAR terminals to receive the message.

5.7.1 Frequency Hopping

To ensure against the eventual development of high-power EHF jamming transmitters. MILSTAR will incorporate an antijamming ploy: frequency hopping. This technique turns a signal into a moving target. Signals transmitted at a given frequency are rapidly shifted, or "hopped," to another frequency. To an enemy, the effect would be like turning to one TV channel only to see the program disappear, then turning to a second channel and seeing that program for a second before it hopped to yet another channel.

MILSTAR frequency-hopping patterns (the sequence and rate of frequency changes) may be changed periodically but not during a particular transmission. The MILSTAR satellite will be able to determine the pattern employed by a specific terminal from its ID code, transmitted at the beginning of a message. Both terminals and satellites will carry clocks to allow transmissions to be synchronized. When several users are transmitting at once, chunks of their signals will zigzag past each other, sharing the system's bandwidth.

This technique is not new either in terrestrial or in satellite applications. Indeed, manual frequency hopping has long been standard practice in military communications. The development of the frequency synthesizer—a unit that determines the carrier frequency electronically—has made automatic frequency hopping practical. England's Racal sells to Third World countries a tactical military radio that automatically changes frequencies, and ITT is developing the SINCGARS automatic hopper for the Army. Some of the transponders on the Navy's FLTSATCOM satellites employ this frequency technique, hopping over a 5-KHz-wide band. Lincoln Laboratories (Lexington, Mass.) has developed a 1500-MHz synthesizer similar in specification to what MILSTAR will require.

New jamming techniques have emerged to counter frequency hopping. One approach is to monitor the spectrum for transmissions and then quickly tune the jamming transmitter to that frequency. Another approach is to jam a narrowband continuously, thereby garbling portions of messages that pass through the band.

To elude sophisticated jamming, MILSTAR transmissions will hop over a broad (1 GHz) band. They will also hop at a high rate (the exact rate is classified).

MILSTAR will carry both data and digitized voice transmissions, all encrypted to protect against eavesdropping. Furthermore, transmission will be accompanied by sophisticated error-correcting codes that will allow a message to be reconstructed even when up to half the bits of information have been erased by hostile jamming or other electronic interference. As an ancillary benefit, such signal processing improves reception by 4 decibels, equivalent to increasing the antenna size threefold.

To allow multiple users to share the satellites, MILSTAR will apply two different multiple access techniques. For the uplink, the system will use frequency division multiple access (FDMA), in which each user is assigned a separate transmission frequency—a technique akin to restricting cars to separate lanes of a highway. Because users will actually be hopping from frequency to frequency to evade jamming signals, strict coordination will be necessary to prevent users from interfering with one another. For the downlink, MILSTAR will use time division multiple access (TDMA), in which user transmissions are interleaved on the channel according to a predetermined schedule. This technique simplifies receiver design, since only one carrier frequency is used for a transmission.

MILSTAR gains its survivability and security features at the cost of performance. Error-correction encoding, encryption, and frequency hopping increase signal bandwidth and hence reduce the satellites' traffic capacity. MILSTAR will handle only 1 megabit of data per second, and no more than 15 users will be able to access a satellite at once. (High-priority users will be able to preempt other users.)

In operation, a MILSTAR satellite would scan the earth looking for users with messages to transmit. The user's message would be digitized and encrypted by the ground station and transmitted to the satellite in the 1 GHz band centered on 44 GHz. The satellite receiver would follow the transmitted signal as it hopped across the band. The satellite's computer would decipher the message to determine its destination and then route it there via a 20 GHz downlink. (The path might include an intermediate relay.)

5.7.2   Trash Can Terminals

The design of earth terminals is not yet finalized; however, according to initial MILSTAR specifications, a typical Army ground station would comprise a transmitter, a receiver, and an antenna—all packaged in a cylindrical container about the size of a trash can. The ground station would be linked by a fiber optic cable to a control unit that would include a teletype keyboard. The station's lid, a highly directional antenna, would pop up and rotate to point at the nearest satellite, whose position would be constantly calculated in the transmitter's memory. The terminal would probably use an offset Cassegrain antenna—a type of antenna that produces a tightly focused beam, minimizing unwanted signal leakage, or sidelobes—to avoid giving away its position to electronic eavesdroppers. The unit would send out a relatively weak (20-watt) signal.

MILSTAR will stretch the state-of-the-art in communications satellites in several key areas. For one thing, MILSTAR satellites will have to do a substantial amount of signal processing in routing messages to their destinations. Present communications satellites are comparatively passive, confining signal processing to frequency conversion and signal amplification. But future civilian satellites will stretch circuit-switching technology even further than MILSTAR, with its small number of users. NASA's Advanced Communications Technology Satellite (ACTS) will use a 20-by-20 intermediate frequency-switch matrix that can handle 100,000 circuits.

5.7.3   Millimeter Wave Technology

Cable television satellites send signals down at 6 GHz, and above, military radar uses about 10 GHz. The EHF band has substantially higher frequencies, i.e., 30 GHz to 300 GHz. Since signals in this band have millimeter wave lengths, the region is also known as the millimeter region. Although interest in this region is growing, relatively little use has been made of it. Certainly no application to date calls for the volume of millimeter hardware that MILSTAR will demand.

Operation in the EHF poses difficult technical challenges, especially for satellite antenna design. Transmission beams at these frequencies are very narrow. Communications with users scattered over the earth's surface will require rapid beam scanning. The magnitude of the problem is illustrated by ACTS, which will also operate in the EHF region.

The four geosynchronous satellites and three inclined-orbit satellites of the MILSTAR satellite network are interconnected by crosslinks as shown in Figure 5.2. The Army, Navy, Air Force, and Marine Corps, users of this network, are depicted. Depending on the scenario, users may be interconnected on the same satellite or via crosslinks to a satellite in another theater. For example, user $U_{1,1}$ may be a command center in CONUS desiring connectivity to a command center in a Navy task force on a Pacific coverage satellite $U_{2, M+K}$.

If the required connectivity must take place in a stressed, jamming, or nuclear environment, there is a requirement to allocate satellite resources and crosslink

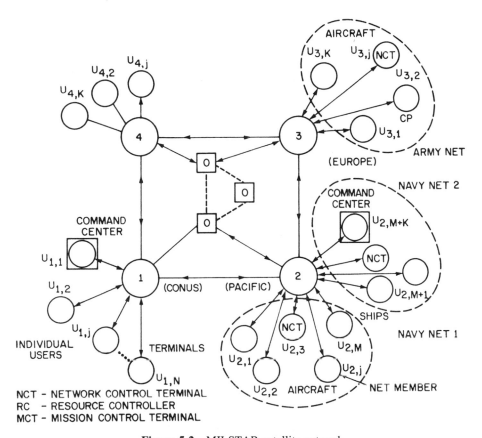

Figure 5.2   MILSTAR satellite network.

capacity very quickly. Furthermore, the satellite must allocate time slots and beams to ensure that the proper secure voice or data requirements are met. In the network, some traffic is intrasatellite (uplink and downlink using only one satellite) and some is intersatellite. In terms of network connectivity, this situation implies that uplinks, crosslinks, and downlinks must be available for communications connectivity and network management under stressful nuclear and jamming situations.

Figures 5.3 and 5.4 show a typical network connectivity for two satellites. Included are the acquisition/tracking, network control (N/C), and access control (A/C) packets and the appropriate protocols for a multipoint network protocol setup. The following definitions pertain to the control packets:

C_0—Data packet
C_1—Control order wire

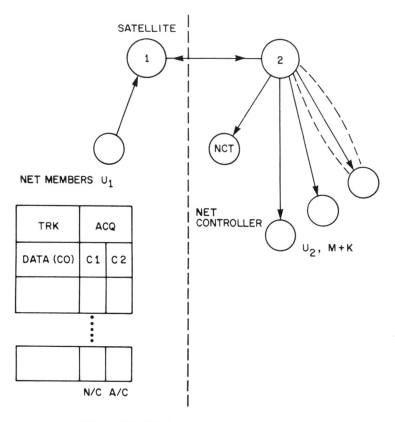

Figure 5.3   Typical network connectivity.

C_2—Access control (uplink)
C_3—Access control (downlink)
C_4—Access control (crosslink)

The MILSTAR system will operate on a demand basis. That is, a user requests service, acquires and tracks the satellite, and via the control access time slots (C_2, C_3 messages) obtains a demand time slot. The process involves a network setup procedure with the MILSTAR resource controller (RC), which assigns time slots and beams to the various terminals as shown in Figure 5.3.

Once all terminals in a network have been assigned, set up, and acknowledged, traffic can flow in the network. Under stressed conditions, network setup and traffic management becomes very critical to the functioning of the forces. Clearly, if it takes too long to set up a network, this may impair the battle situation.

Because of the classified nature of the MILSTAR system, details regarding controls and signal structure cannot be elaborated upon. As the classification status of MILSTAR changes, additional data will be made available.

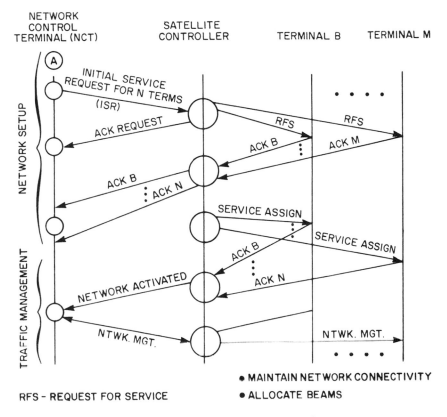

Figure 5.4 Typical network protocol.

5.8 AFSATCOM SATELLITE SYSTEM

AFSATCOM I, also known as FLTSAT (see Chapter 4, Section 4.9), is operated in the UHF band only and consists of 12 narrowband channels (5 KHz center frequency separation) and one wideband channel with 500 KHz bandwidth. The uplink narrowband channel has three operational modes: (1) normal mode—12 fixed-frequency channels; (2) alternate mode—7 channels frequency shifted, 5 channel frequencies remain at normal; (3) stressed mode—7 channels frequency hopped in group, 5 channel frequencies remain at normal. Narrowband channel mode selection is by UHF command without affecting wideband channel operation. The 12 NB downlink channels utilize spread-spectrum modulation. The data modulation on the AFSATCOM I WB channels is binary FSK with mark space tones separated ± 1.25 KHz from carrier. The wideband channel is a code division multiple access channel. The data modulation could be FSK and depends on an end-to-end user.

The AFSATCOM II signal design employs both UHF and SHF bands for uplink and downlink communications. In addition, a beacon operated at SHF is included to facilitate SHF signal sync-acquisition. Aside from the differences in carrier frequency and spread-spectrum bandwidth, the signal structures for UHF and SHF links of AFSATCOM II are essentially the same. For the single channel transponder (SCT) onboard Satellite Defense System (SDS), DSCS III and Defense Satellite Program (DSP), the single uplink channel signal can be either at fixed frequency or hopped over 64 MHz for UHF and 48 MHz, 102.3 MHz, or 409.6 MHz for SHF. The downlink hop bandwidth is either 800 KHz or 6.4 MHz for both UHF and SHF links. The 8-ary FSK modulation for the uplink uses chip-combining techniques; however, for the downlink signal, Queen's code is used to convert the 8-ary tones to data bits. There is also an alternate modulation design for the AFSATCOM II uplink/AFSATCOM I downlink mode.

Additional information on the FLTSAT can be obtained from DoD documentation. The classified nature of this material makes it difficult to include any additional details.

5.9   DEFENSE SATELLITE COMMUNICATIONS (DSCS)

The Defense Satellite Communication System (DSCS) is presently utilizing the DSCS II system. In the future, the DSCS III, post-DSCS III, and commercial satellites will be utilized to a greater extent. In terms of AJ capability and wartime functioning, the DSCS will play a significant role. Detailed descriptions of the DSCS can be found in the open literature. The element of the DSCS that is of relevance to the military is its electronic counter-counter-measure (ECCM) net and its control system. The control aspects will be treated in more detail later in this chapter.

The DSCS III communications satellite is being developed and is undergoing in-orbit testing as a replacement for the current DSCS II space segment. Key objectives of the DSCS III are (1) improved life cycle costs by means of longer design life; (2) significant improvements over the DSCS II in its ability to withstand electronic warfare environments, mainly by its multibeam antenna (MBA) system; and (3) increased flexibility to provide optimum service for an increasing and dissimilar set of user groups.

The most significant operational features of the DSCS III are in the areas of electronic counter-countermeasures (ECCM) and flexibility of operation. Those capabilities are made possible by the antennas onboard the DSCS III spacecraft, as shown in Figure 5.5. The flexibility to contour coverage areas to match user locations and gain requirements is provided by a 61-element receive MBA and two 19-element transmit MBAs. The receive MBA affords dramatic mitigation against jammers by providing the capability for detecting, locating, and placing a null on jammers while leaving the gain on user terminals relatively undisturbed. Nulling is not constrained to simple beam turnoff but utilizes phase and amplitude control of all 61 feed horns to enhance its ability to discriminate against jammers.

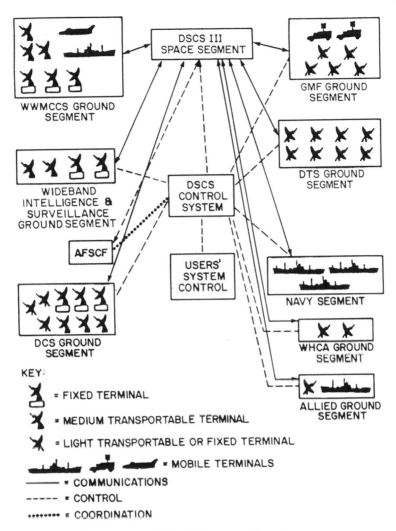

KEY:

= FIXED TERMINAL

= MEDIUM TRANSPORTABLE TERMINAL

= LIGHT TRANSPORTABLE OR FIXED TERMINAL

= MOBILE TERMINALS

——— = COMMUNICATIONS

----- = CONTROL

········· = COORDINATION

Figure 5.5 Major DSCS elements.

This capability is accomplished by an advanced design lens. The receive beam-forming network (BFN) is composed of 121 variable power dividers (VPDs) and phase shifters that allow precise and incremental control of the combination of the signals from each of the individual feed horns.

Full attainment of the benefits of the flexibility of the MBAs, other antennas, and the multiple transponders of the DSCS III spacecraft is ensured by the development of a real-time, adaptive-control system (RTACS). The RTACS will provide computer-assigned automation of status gathering, computation of user demands, and resource allocation on a priority basis; assistance in the collection and analysis of status information; analysis of changes in user requirements; and generation and optimization of alternative system configurations.

The predominant users of the DSCS are (1) the Worldwide Military Command and Control System (WWMCCS), (2) ground mobile forces, (3) Navy ships, (4) Wide Band data relay, (5) the Defense Communications System (AUTO-VON, AUTODIN, etc.), (6) White House communications, (7) the Diplomatic Telecommunications System (DTS), and (8) systems providing support to allied nations. When the DSCS III reaches full operational capability, it is expected to serve approximately 400 earth terminals. The current system is supporting over 130 earth terminals.

Support for these users is planned via worldwide deployment of four operational satellites located at longitudes of 12°W, 135°W, 175°E, and 65°E. High availability will be ensured by maintaining two spare satellites in orbit.

5.9.1 Summary of Key DSCS III Features

The DSCS III design provides a flexible configuration to allow optimization of the transponder transfer characteristics for each major grouping of users. Likewise, the antenna coverage of four of the transponders may be contoured to concentrate power and G/T only to those areas of interest. This arrangement not only enhances the antenna gain but excludes areas of no interest, greatly reducing the potential for interference with or by other users.

Table 5.4 lists the salient features of the DSCS III. These features allow the tailoring of the spacecraft configuration to best meet the users' requirements within the constraints of a given scenario. The frequency plan of Figure 5.6, developed in 1974, was constrained by the International Telecommunications Union regulations of that period. Notable in this design is the dual frequency translation that is required if (1) the fixed satellite "exclusive" band (no sharing of frequency allocations with terrestrial services) is to map on both uplink and downlink frequencies and (2) full use of the 500 MHz allocation is desired. This

Table 5.4 Major features of the DSCS III.

- Six independent transponders
  - Each with its own traveling wave tube amplifier (TWTA)
    - two 40 watt TWTAs
    - four 10 watt TWTAs
- 61-element receive multibeam antenna and beam-forming network, which allow:
  - selective coverage
  - jammer detection and location
  - multiple jammer nulling
- Two 19-element transmit antennas, which allow:
  - selective coverage for Ch. 1 and 3

- selective coverage for Ch. 2 and 4
- A gimballed dish antenna, which:
  - provides high EIRP on an area-coverage basis
  - may be connected to transponders 1, 2, or 4; or 1 and 4; or 2 and 4
- Earth-coverage receive horns for transponders 5 and 6 on a full-time basis and transponders 1 through 4 selectively
- Earth-coverage transmit horns for transponders 5 and 6 on a full-time basis and transponders 3 and 4 selectively
- X-band TT&C with protection against jamming
- Single-channel transponder—provides AJ protected uplink at SHF and UHF with UHF downlink

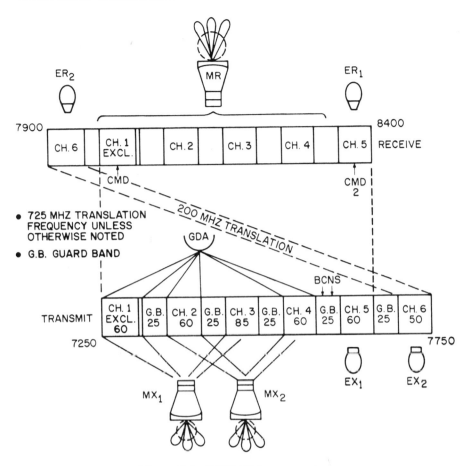

Figure 5.6 DSCS III frequency plan.

dual translation was required since the ITU "exclusive" allocations had a 725 MHz offset, whereas the full frequency band offset is 650 MHz.

Authorization of satellite communications to and from mobile platforms has been granted over the lower 125 MHz of the uplink and downlink frequencies, a 650 MHz offset, which overlaps the exclusive band. This presents a conflict between aligning the mobile authorization and aligning the exclusive band. Additional actions at WARC-79 reduced the protection of the exclusive band from a table entry to a footnote status, providing protection to satellite services only within those nations listed in the footnote. Design changes to the DSCS III in response to the revision in allocations will be analyzed for possible inclusion in the second block of production satellites. Other changes to the DSCS III are being considered on a product improvement basis as component technology is improved. One area under current investigation is that of substituting solid-state amplifiers (SSAs) for the current TWTA power output devices, with the goal of a significant improvement in reliability. Any changeover to SSAs would be on an incremental basis rather than a total changeover to SSAs.

Careful analysis of Figure 5.6 reveals that 16 different receive antenna configurations and 24 potential transmit antenna configurations are possible. A typical configuration will utilize the receive MBA on transponders 1 through 4 and earth coverage receive antennas on 5 and 6. A corresponding transmit connectivity may utilize transmit MBA No. 1 on transponders 1 and 3, the gimballed dish antenna (GDA) on transponder 2, Earth Coverage (EC) horns on transponders 5 and 6, leaving transmit MBA No. 2 to be configured according to the needs of the users of transponder 4.

While the multiple transponders and flexible antenna connectivity allow for tailoring the spacecraft communications package to meet changing user requirements and the demands of four orbital locations, this same flexibility and the selective tailoring of the MBA receive and transmit antenna patterns pose a significant challenge to achieving optimum allocation of spacecraft resources on a dynamic basis—especially in responding to attacks by jammers. The following sections will review several programs underway to facilitate adaption and optimum resource allocation.

5.9.2 ECCM Net

The DSCS ECCM net is provided via a 60 MHz wideband channel at SHF, for both uplink and downlink configurations, as shown in Table 5.3. It utilizes direct sequence, pseudonoise (PN) code modulation for band spreading. The PN chip rate varies from 10 M chips per sec to 40 M chips per sec, depending on the end-to-end users. The multiple access employs both code division multiple access (CDMA) and time-division multiple access (TDMA) techniques with frame rate at 4800 bits per frame and time slot rate at 256 bit per sec. Quadriphase shift keying (QPSK) is used for data modulation with data rates from a low of 75 b/s to a high of 2.5 mb/s.

5.9.3 ECCM Net or Wideband CINCNET Channel Processing Requirements

The DSCS ECCM net or the wideband CINCNET channel of FLTSATCOM is for user-to-user direct communication without any signal processing on board the satellite. Processing functions required at the terminal for ECCM net or wideband channel thus depend on users' operational needs.

5.10 WIDEBAND/NARROWBAND USE OF DSCS/MILSTAR

Although future satellite requirements have not been thoroughly defined, it is assumed that both DSCS and MILSTAR will accommodate low data-rate users.

MILSTAR will accommodate users at rates of 75 bps to 2.4 kbps with a few 16 kbps users. The future DSCS system will accommodate data rates from 75 bps to 4.6 Mbps. Some, but not all, of the lower data-rate users (75 bps–16 Kbps) on DSCS may also be able to operate through or with MILSTAR. The requirements for future DSCS users are similar to the requirements motivating MILSTAR development, but inherent capacity limitations of MILSTAR lead to a need for a demand access capability on DSCS as well.

The development of a control concept for future DSCS depends very much on the defined Army, Navy, Air Force, and Marine Corps requirements. It is assumed that requirements for tactical users are consistent with those outlined for the ground mobile forces (GMF). For the strategic users, it is assumed the requirements are related to the Tactical Warning/Attack Assessment (TW/AA) and Command Centers (CC) operational scenarios (note Figure 5.7).

Future DSCS requirements will have to satisfy wideband Army, Air Force, and Marine Corps users utilizing multichannel 32 Kbps CVSD up to 144 channels in addition to wideband EHF users active in the post Phase III era. Figure 5.8 shows the MILSTAR and DSCS system that could accommodate strategic and tactical users in the future.

Multichannel satellites provide wideband connectivity for voice and data at the TRI-TAC large switch nodes while the single channel (S/C) satellites (16 Kbps) provide connectivity between echelons for data communications of the mobile subscriber (MSC) terminals and transportable command centers. Multichannel terminals (M/C) will communicate with other M/C terminals within the same echelon or other echelons. Single channel (S/C) terminals will communicate with other S/C terminals for data transmission but will not communicate with M/C terminals. The Air Force and Marine Corps intend to use satellites in a similar operational environment. The key consideration is that tactical users can require more than one 16 Kbps link in a battlefield situation. Strategic users may also require more than one 16 Kbps channel, but these requirements have not been explicitly defined.

Development of a DSCS EHF demand-assignment multiple-access (DAMA) capability can benefit from the significant R&D effort already invested in MIL-STAR and the Navy UHF DAMA program. The underlying narrowband user requirements for DSCS DAMA are parallel in many respects to the user requirements for MILSTAR. Building on the heritage of MILSTAR, eventual R&D and deployment costs can be held down and potential obstacles to DSCS/MILSTAR interoperability avoided.

While many DSCS functions parallel those of MILSTAR, some differences exist. The satellite constellations differ, with DSCS employing nearly geostationary orbits and MILSTAR employing a mix of nearly geostationary orbits and inclined orbits. Crosslinking patterns are also different; DSCS is currently envisioned as employing a ring topology with two crosslink ports per satellite, while MILSTAR, using a greater number of satellites, employs an adaptive routing scheme. These differences account for a more simplified crosslink routing control problem in DSCS than in MILSTAR. Furthermore, there are differences

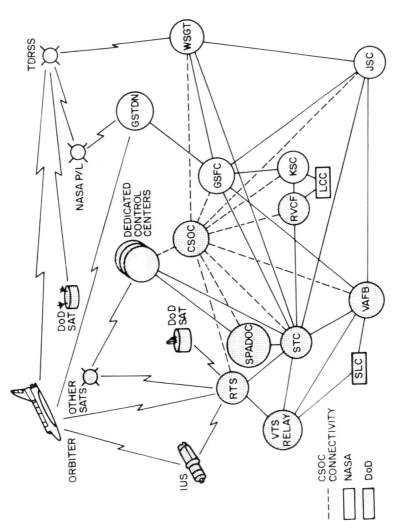

Figure 5.7 Space Operations Network—CSOC timeframe.

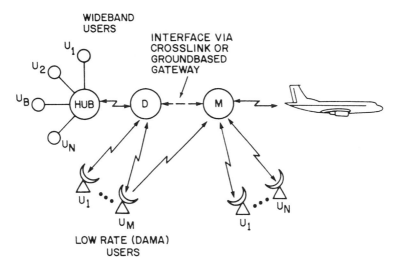

Figure 5.8   Future DSCS/MILSTAR user environment.

in communication requirements. MILSTAR is to accommodate low rate 75 bps to 2.4 kbps voice and data links in a nuclear environment. DSCS will support low rate, as well as higher rate, users for both peacetime and wartime situations. At the high end of the data-rate spectrum, a MILSTAR terminal can support a single 16 kbps data channel or up to four separate channels at 2.4 kbps each (in this case, control overhead is added by the terminal). Certain DSCS terminals may support multiplexed groupings of up to 144 channels, 32 Kbps each. These large multiple groupings are associated with the TRI-TAC family of switch equipment. Their high-bulk data rates, up to 4.6 Mbps, preclude the utilization of MILSTAR-derived equipment at the high end of the data spectrum. Under development is the Sat Data Link Standard to accommodate this.

5.11   A STATE-OF-THE-ART SATELLITE SYSTEM

Generally, the future satellite communications systems will consist of smaller terminals, intelligent onboard processing, sophisticated signal processing techniques and more efficient modulation schemes. Figure 5.9 shows an example of an end-to-end state-of-the-art system utilizing a time division multiple access (TDMA) scheme and crosslinks to other satellites to provide for robust digital communications. More specifics on digital satellite communication can be obtained in [1].

The TDMA scheme that will be utilized in the future is a multiple-access technique that permits individual earth terminal transmissions to be received by

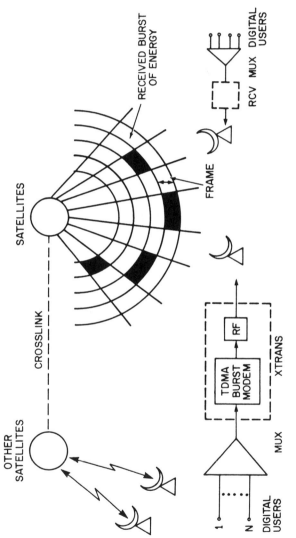

Figure 5.9   State-of-the-art TDMA satellite system.

the satellite in separate nonoverlapping time slots, thereby avoiding the generation of intermodulation products in a nonlinear transponder. Each ground terminal must determine satellite system time and range so that the transmitted signals are timed to arrive at the satellite in the proper time slots. As shown in the figure, each high-velocity burst of RF energy, typically quadriphase (QPSK) modulation, arrives at the satellite in its assigned time slot.

Time-division multiple access (TDMA) permits the output amplifier to be operated in full saturation, often resulting in a significant increase in useful power output (as contrasted with 3 db backoff). Intermodulation (IM)-product degradation is largely avoided by transmitting each signal with sufficient guard time between time slots to accommodate any timing inaccuracies, while preventing the "tails" of the pulsed previous and next signal from causing significant interferences in the present time slot. The amplitude of these tails depends on the transient response and, in turn, on the amplitude and phase response of the filters, both in the satellite transponder receive section and in the earth terminal transmit filters.

If the transponder is operated in the "hard-limiting" mode and limits on noise input alone, the output envelope is essentially constant, even during the guard-time interval. Typically, guard times can be made sufficiently small that the total guard-time frame consumes less than 10 percent of the usable signal power and the transponder is utilized to greater than 90 percent efficiency.

Each TDMA earth station has parallel input digital bit streams, or analog streams that are digitized at the earth station, which are addressed to separate receiving earth stations. The signals addressed to separate receive terminals are allocated separate portions of the transmit TDMA burst following the TDMA burst from separate transmit terminals and multiplexers. The appropriate portions of them are then demultiplexed into separate serial bit streams.

TDMA system timing is such that if all earth stations transmitted at the beginning (epoch) of their respective frames, all signals would arrive simultaneously at the satellite. If the frame rate is $f_f = 1/T_f$, all input data rates f_{di} must be exact integral multiples of f_f—that is, $f_{di} = n_i f_f$. Otherwise, an integral number of bits could be transmitted during each frame (or superframe). The burst rate f_{bi} is usually integrally related to the frame rate because $f_{di} T_f$ data bits are in each burst, and each burst duration is a natural fraction of the frame duration. Ordinarily, the burst rate should be the largest rate permitted by the satellite ERP and the G/T (antenna gain/noise temperature) of the receiving ground station. If desired, the burst rate in one portion of the burst from a given terminal can differ from that used in other portions by some multiple of the frame rate.

Parallel voice channels are PCM encoded at a clock rate synchronous with the TDMA frame rate. If there is a multiplicity of voice channels at the particular terminal, then this PCM technique can take advantage of the loading factor of voice, which is less than 50 percent because of pauses in speech. Voice activity can be sensed and channel sharing used for the active channels as in Time-Assigned Speech Interpolation (TASI). Thus, benefits similar to those for single-channel-per-carrier demand-assigned-multiple-access (SCPC-DAMA) can be obtained in addition to the other TDMA advantages.

Parallel PCM bit streams must be multiplexed together at a rate synchronous with the TDMA frame rate. Thus, pulse-stuffing multiplexing or elastic buffers usually must be employed to synchronize the bit streams. The coded serial bit streams are then fed to compression buffers, where bits stored during one frame are burst out in the appropriate time slot. Frame timing is controlled by a separate timing unit, which may utilize the initial portion of the frame for ranging/timing transmissions. Timing within the frame, or within a TDMA burst, is controlled by the synchronization burst generator and synch-burst control unit.

5.12 CONTROL OF SATELLITE SYSTEMS

The control of satellite resources, especially in a hostile environment, involves an intricate combination of coordination among the DoD hierarchy of command including the National Command Authority (NCA) and related European, Pacific, and Indian Ocean command authorities. With the advent of the Space Command, additional functions must be considered for the proper control of satellite resources. This section will highlight some of the critical control issues and techniques as well discuss a new approach to control and the consolidated Space Operations Center (CSOC).

The design and control of military satellite communications (MILSATCOM) involves a mix of operational and technical issues. Because of their inherent global nature, satellite systems are a somewhat unique communications asset. Operationally, MILSATCOM systems must be managed as major resources of the Defense Communications System (DCS). Technically, MILSATCOM systems must support end-to-end user connectivity in a potentially fluid, dynamically changing environment. In a warfighting situation, this may include physical destruction of major system assets. The newest element of satellite control facility is the CSOC.

5.12.1 Consolidated Space Operations Center (CSOC)

When CSOC is fully operational and the data system modernization (DSM) program of the Air Force Satellite Control Facility (AFSCF) has been fully implemented, CSOC will function as an operational center within the Air Force Space Control Network (AFSCN).[1] This system definition refers to the AFSCN operational era. As such, the configurations and operations described herein represent the expansion, consolidation, and internetting of current and planned U.S. Air Force telemetry, tracking, and command (TT&C) capabilities.

---

[1]The information contained in this section was obtained from the Air Force specification for CSOC.

The AFSCN is a worldwide configuration of space-ground link TT&C resources interconnected through communications links to data processing control centers. Ground and space resources comprising this network are managed, operated, and supported by trained military, Department of Defense (DoD) civilian, and contractor personnel. CSOC is located at Falcon Air Force Station near Colorado Springs, Colorado. The three basic elements of the AFSCN, shown in Figure 5.10 (remote tracking stations (RTS), communications, and control centers), interface with space vehicles, with DoD and the National Aeronautics and Space Administration (NASA) network. The AFSCN-user vehicles may include DoD and non-DoD satellites in all phases of operation, the Space Shuttle Vehicle (SSV) for DoD missions and for NASA missions when certain emergency conditions exist, and ballistic missiles (with occasional telemetry support of test launches).

The major DoD resources that are tied to AFSCN assets include the Space Defense Operations Center (SPADOC), the Eastern and Western Space and Missile Centers (ESMC/WSMC), and various users of mission data. SPADOC, which is responsible for maintaining the Space Order of Battle, provides the military forces assigned to organizations that operate AFSCN space elements and directs the execution of defensive and offensive space operations. CSOC reports the operational status of assigned ground and space assets to SPADOC. ESMC/WSMC are linked to CSOC and other AFSCN resources to facilitate integrated prelaunch space vehicle checkout and to coordinate launch operations.

NASA resources used to support AFSCN operations include the Johnson Space Center (JSC), the Kennedy Space Center (KSC), the Goddard Space Flight Center (GSFC), the White Sands Ground Terminal (WSGT), the Tracking and Data Relay Satellite System (TDRSS), the Space Tracking and Data Network (STDN) ground systems as well as the Space Shuttle Vehicle (SSV) itself.

CSOC is a secure facility dedicated to the planning and control of DoD space missions. It includes a satellite operations complex (SOC) and a shuttle operations and planning complex (SOPC), plus required capabilities to support SOC, SOPC, and collocated program elements (CPEs). The Global Positioning System Element (GPS/E), consisting of a master control station (MCS) and a monitor station, is the only CPE at this time. The Colorado Tracking Station, which is part of the AFSCF chain of remote tracking stations, is also collocated with CSOC.

The fully operational CSOC assumes primary responsibility for planning, readiness, and control operations and support for all DoD shuttle missions and assigned satellite programs. CSOC system capabilities will be implemented incrementally and interfaced with existing and new elements of the AFSCN and other external systems.

5.12.1.1 SOC Segment

SOC assumes primary responsibility for the preflight, flight, and postflight operations necessary to satisfy DoD satellite mission objectives by using computer-and-display and control equipment, software, and procedures. SOC is the

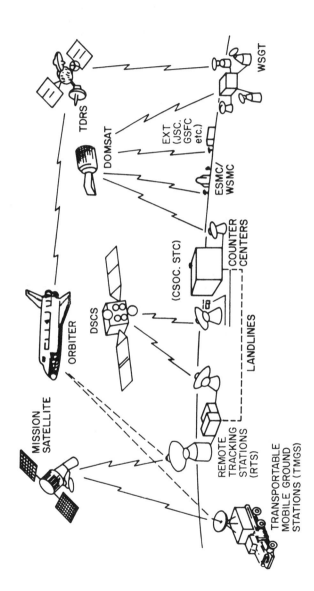

Figure 5.10 The AFSCN and supporting resources.

STC – SATELLITE TEST CENTER
TDRS – TRACKING DATA RELAY SYSTEM
WSGT – WHITE SANDS GROUND TERMINAL
ESMC – EASTERN SPACE MISSILE CENTER
GSFC – GODDARD SPACE FLIGHT CENTER

physical and functional equivalent of portions of the AFSCF Command and Control Segment (CCS), as developed and acquired by the data systems modernization (DSM) program. SOC interfaces with SOPC to plan, schedule, and prepare payloads to be launched from the space shuttle vehicle (SSV), to operate payloads within the SSV (sortie missions), to deploy upper stages and free-flyer satellites, and to retrieve payloads from orbit.

5.12.1.2 SOPC Segment

SOPC assumes primary responsibility for the preflight, flight, and postflight operations necessary to satisfy DoD SSV mission objectives. SOPC includes computer and display equipment, data distribution equipment, software, and procedures to perform shuttle flight operations, planning, and other assigned functions.

SOPC also supports the planning, scheduling, and preparation of DoD payloads to be carried on the SSV, the operation of payloads within the SSV, deployment of upper stages or free-flyer satellites, and retrieval of payloads from orbit.

5.12.1.3 Communications Segment (CS)

The CS consists of the communications circuits and equipment, e.g., multiplexers and demultiplexers, switching and routing equipment, modems, recorders, communications security (COMSEC) equipment, and preprocessors. The CS supports command, telemetry, tracking voice, video, teletype, facsimile, and other data communications and data services internal to COSC and externally among CSOC, other AFSCN resources, the NASA communications (NASCOM) network, the spaceflight tracking and data network (STDN), CPEs, and other government communications services and facilities. In addition to the equipment located at the Colorado site, the CS is defined to also contain interface equipment located at the remote ends of several of the AFSCN and NASCOM links.

5.12.1.4 Facilities Segment (FS)

The FS consists of all centralized real-property installed equipment, grounds, buildings, utilities, grounding, environmental controls, storage, perimeter fencing and lighting, roadways, parking, and accommodations required for CSOC and CPEs.

5.12.1.5 Support Segment (SS)

The SS consists of an uninterruptible power supply (UPS), a timing subsystem (TS), a weather support unit (WSU), a technical data center (TDC), a security control subsystem (SCS), and an operations command center (OCC).

5.12.1.6   Validation Segment (VS)

The VS validates the CSOC mission and configuration and trains and certifies personnel to support specific shuttle and spacecraft missions. The VS comprises the CSOC Simulation System (CSS) and the Training Equipment System (TES).

5.12.1.7   Network Control Segment (NCS)

The NCS consists of hardware and software required to schedule AFSCN and NASA resources necessary for support of CSOC/AFSCN operations and to control AFSCN resources. The NCS includes a capability that functionally duplicates and is interoperable with the range control complex within the Satellite Test Center (STC), which is also an element of the AFSCN.

The NCS supports the SOC and SOPC, by scheduling and allocating tracking resources on request, including those of TDRSS through cooperation with the NASA Goddard Space Flight Center (GSFC). NCS also supports CSOC by scheduling other CSOC resources. The SS supports the other segments by providing uninterrupted power, time signals, weather data, technical library services, a variety of management support services, as well as physical security. The Operations Command Center (OCC) provides direction to other segments in emergency situations. Finally, all communication support required by the segments internal to CSOC is provided by the CS.

5.12.2   External Interfaces

External interfaces exist with other Air Force and DoD facilities and with NASA and other non-DoD facilities, as indicated in Figure 5.7.

SOC interfaces with the STC, the remote tracking stations (RTSs), and other AFSCF elements to support satellite missions. SOC also interfaces with elements of the NASA network including JSC, GSFC, WSGT, and KSC.

SOPC interfaces with elements of the NASA networks, including the Johnson Space Center (JSC), the Goddard Space Flight Center (GSFC), the Kennedy Space Center (KSC), and the White Sands Ground Terminal (WSGT) of TDRSS, for support of space shuttle missions. SOPC also interfaces with STC to provide RTS communication and tracking backup in the event of TDRSS failure, and to supplement TDRSS tracking.

The CS provides communication interfaces with various external agencies to support SOC and SOPC data exchanges for planning, readiness, and flight control. CS also interfaces with CPEs to support CSOC and CPE operations and to coordinate shared use of resources.

The NCS interfaces with the STC's RCC for interoperable scheduling and control of DoD resources and with NASA resource scheduling functions for requesting scheduled use of NASA resources. Figure 5.11 presents an overview of TT&C ground resources that the NCS schedules.

The VS interfaces with the STC to provide simulated spacecraft data for integrated joint training exercises with the STC MCCs. It interfaces with JSC

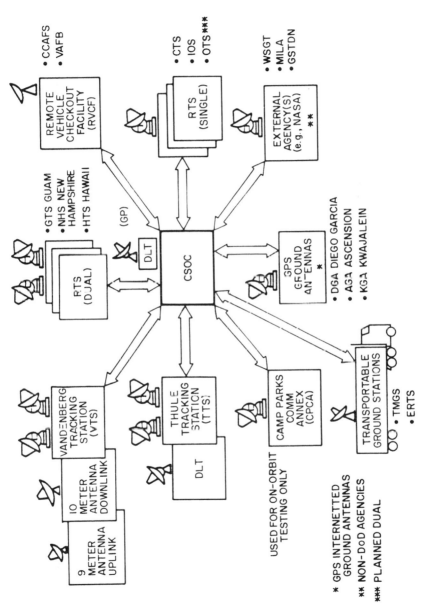

Figure 5.11 TT&C ground resources.

to provide simulated SSV data for integrated joint training exercises with JSC controlled-mode flight center rooms (FCRs). The VS also receives JSC-generated SMS data for use in training exercises.

The SS Weather Support Unit (WSU) interfaces with DoD and other government weather services to obtain raw and processed weather data for use by other segments. The Operations Command Center (OCC) of the SS interfaces with higher military command centers to receive directives to be implemented by the segments. The timing subsystem (TS) of the SS receives external timing inputs and the technical data center (TDC) provides technical and management inputs.

The issues of system management, communication protocols, and control system architecture become increasingly interdependent as the trend toward automation continues. For example, the decision to employ satellite on-board traffic control affects the functions of ground-based DCS control facilities. Similarly, as automated network functions become more complex, and user requirements become more exacting, the underlying user-to-network protocols become increasingly sophisticated.

This section describes, in turn, each model of Figure 5.12 illustrating viewpoints of control of communications and highlighting the interplay between models. Their application to satellite communication systems is emphasized. Examples are taken from the Defense Satellite Communications System (DSCS)[2] and MILSTAR.[3] In addition, a look to the future examines the trend toward further integration as systems become more capable and sophisticated.

5.12.3   DCS System Control

Figure 5.13(a) illustrates system control as described within the DCS. It consists of a functional classification into four major areas and a hierarchical framework consistent with military operations.

5.12.3.1   Control Functions

System control functions are identified as (1) network control; (2) traffic and routing control; (3) performance assessment and status monitoring; and (4) technical control, patch and test. Network control is primarily concerned with the topology of a communication system under normal and stressed conditions. It includes the original and extended system design (its topology), protocols for responding to jamming and physical attack, and assignment of channel blocks to user subnetworks and military services. The original design, including definition of control algorithms and planning for system extensions, are background functions.

---

[2]The DSCS is discussed in terms of the DSCS III Real Time Adaptive Control System (RTACS) era. As it evolves in the future system control during this time period and beyond will fully integrate satellite, earth station, and network control into a single structure, with direct control and management responsibility to the appropriate levels of the DCS control structure [3].

[3]MILSTAR is one of the planned ''new start'' military SATCOM systems.

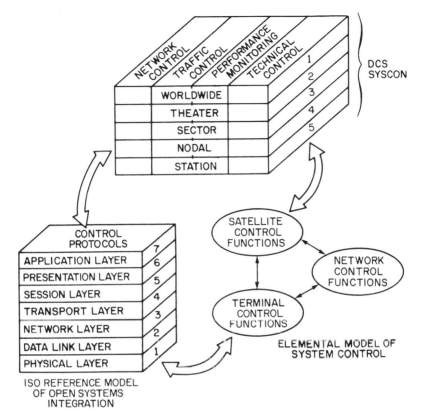

Figure 5.12 Viewpoints on control of communications.

Channel assignment and application of specific recover protocols involves real-time as well as background functions.

Traffic and routing control is primarily concerned with flow control, congestion management, and channel assignment in a switched network. In SATCOM systems with limited and expensive communication resources, this generally implies some form of demand assignment. Historically, routing control in SATCOM systems has relied on terrestrial switching and relay functions. This was a consequence of essentially disconnected satellite constellations, where satellites communicated with numerous ground stations but not with each other. Current and future SATCOM systems will incorporate satellite crosslinking—direct communications between satellites. This change will allow worldwide routing without ground-based relay. MILSTAR will be a major advance in this direction, incorporating dynamic crosslinks between satellites in similar and dissimilar orbits. Crosslink control and routing of user data throughout the satellite constellation will be accomplished automatically by the satellites themselves. However, the control algorithms will be traceable to the theater and worldwide levels of the DCS System Control (SYSCON) hierarchy.

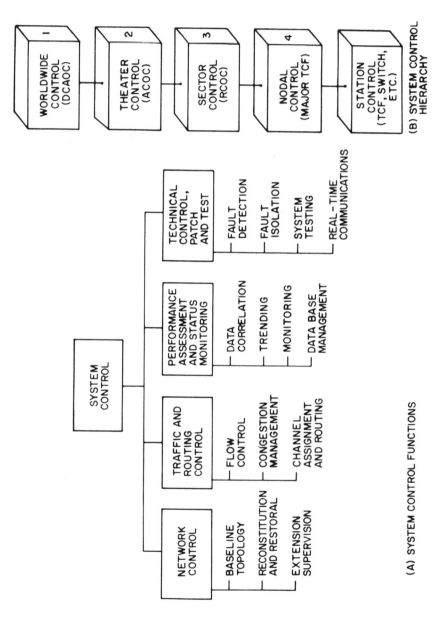

Figure 5.13 DCS system control.

Performance assessment and status monitoring provide data on user equipment and network facilities and assessments of current and historical system health. This data supports management review of system performance, rapid response to changing conditions, and cost-effective logistical support, maintenance, and testing of system elements. The status monitoring function accepts and processes status data collected throughout the system and reported to a central facility. Performance assessment extracts management information from this data base, including current networkwide system performance and performance trends. These functions are generally centralized. Centralization reduces channel overhead and avoids the requirement for sophisticated computing facilities at all nodes in the SATCOM system. In the Defense Satellite Communication System's Operational Control System (DOCS), described below, this function is supported at the worldwide and theater levels, with theaterwide analysis performed at the RTACS Network Control Facility (NCF).

Technical control, patch, and test are local functions performed at all network telecommunication facilities, including active in-orbit spacecraft. This area includes quality assurance, monitoring, patch, test, coordination, restoral, and reporting. These tasks support technical supervision and control over trunks and circuits traversing or terminating in a facility. The functions complement and support performance assessment and status monitoring. The status data reported to a central monitoring facility is collected as part of the technical control, patch, and test objective area. In regard to SATCOM systems, these functions are performed at all ground stations and on board the satellite(s).

5.12.3.2 Control Hierarchy

Turning now to the system control hierarchy illustrated in Figure 2.13(b), five levels are defined and configured to "provide for the lowest level of restoral and reconfiguration execution and the highest level of control." [4]

This doctrine enhances DCS survivability and ensures visibility. In this context, "control" implies background management and planning. The objectives are (1) to maximize worldwide communications efficiency by performing configuration control, extension supervision, traffic and routing policy development, and performance assessment at the highest practical level, and (2) to maximize survivability by responding to stress and restoring all available service, from the lowest practical level.

The first two levels, worldwide and theater, are staffed and operated by the Defense Communications Agency (DCA). Worldwide control is exercised at the DCA Operations Center (DCAOC). Theater control is exercised by two area communications operations centers (ACOCs), one each for the Atlantic and Pacific. The Pacific ACOC is supported by several regional communications operations centers (RCOCs), which perform operational direction of the various transmission media. Each level coordinates with users and services at its own level. The White House Communications Agency, JCS, military services, and other users with global activity coordinate through the DCAOC. Theater-level commands and services coordinate through the ACOCs and RCOCs. Each hierarchical level directs the activity of subordinate elements and coordinates activity between subordinate elements.

A significant advance, made possible by the ongoing automation and integration of DCS control functions, is the ability to extend the fault isolation function up to the theater level. Correlation functions applied at this level can be used to detect and isolate failures and identify adverse trends [5] that would be difficult to identify from a local perspective. In the DSCS RTACS, this function is performed by the Network Control Facility (NCF) at the equivalent of the theater level (see Figure 5.14). The NCF interfaces with worldwide and theater-level operations centers through the Operations Control Element (OCE). In the future MILSTAR system, discussed below, many of the traditional real-time functions will be performed on board the satellite. However, background control and management will still be exercised by human operators at the worldwide and theater level.

The sector, nodal, and station levels of the control hierarchy consist of numerous facilities operated and staffed by the various military services. These facilities coordinate with users at their own hierarchical level, direct the activity of subordinate elements, and coordinate activity between subordinate elements. In addition, these facilities perform technical control, patch and test, and local performance monitoring and assessment. SATCOM systems do not differentiate between these hierarchical levels from a telecommunications standpoint. The single ground telecommunication facility is a terminal. In the DSCS, each ground terminal is associated with a terminal control element (TCE). The TCE interfaces with traditional station-level and nodal-level control facilities in order to conform with existing reporting and command channels. Intentional and unintentional changes in satellite performance and connectivity are therefore monitored simultaneously via the RTACS hierarchy and through the usual control monitoring hierarchy. This structure, with its multiple reporting paths, yields a high degree of flexibility, reliability, and survivability [3].

5.12.4 Control of Satellite Resources

The control of satellites involves network control, traffic control, technical control, and performance control. Under stressed conditions, the various aspects of control must be utilized to provide for unimpaired end-to-end connectivity. Although control systems have been designed and are in operation for the current DSCS system, there appears to be a need for a new look at the control problem, especially with the proliferation of satellite systems.

The control system must provide for communications support, system survivability, and reconfiguration under stress; yet it must be technically and economically efficient. To achieve these multiple goals, the overall control architecture must assign each control function to the appropriate physical element of the MILSATCOM system. The final control architecture will differ from system to

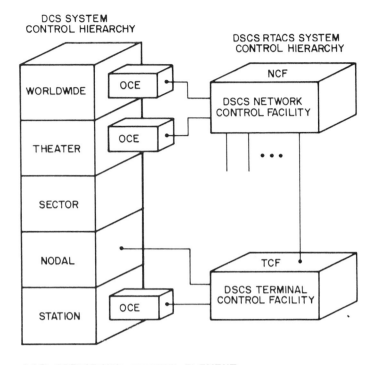

OCE: OPERATIONAL CONTROL ELEMENT
NCF: NETWORK CONTROL FACILITY (INCLUDES THE NETWORK
CONTROL ELEMENT—NCE)
TCE : TERMINAL CONTROL ELEMENT

Each level within the DCS hierarchy performs these generic functions:

1. Coordination with users and services at the specified level (e.g., the worldwide level interfaces with the White House Communications Agency—WHCA). This function includes status reporting and user advisories.
2. Coordination of communications among subordinate DCS elements or regions.
3. Direction of subordinate elements.
4. Upward reporting to higher levels.
5. Functional elements of system control (Network Control, Traffic Control, Performance Assessment and Status Monitoring, and Technical Control, Patch and Test) as required at the specified level.

Figure 5.14 DOCS interfaces.

system, depending on user requirements, technical state-of-the-art, cost constraints, and existing communications facilities that require access to the new system.

The solution to these technical issues depends on standard communication protocols and a cost-effective control system. Standard protocols allow end-to-end connectivity within arbitrary groups of users. Standard protocols also allow the system to accommodate new users without continual and expensive redesign.

The three issues—system management, communication protocols, and control system architecture—can each be understood through the use of a model. No single model offers an equally clear view of all three issues, but each issue is clearly illustrated with its own model. Figure 5.12 illustrates three models in widespread use today:

1. DCS SYSCON—The Defense Communications System's definition of System Control, DCS SYSCON, highlights the management structure of military communications systems.
2. ISO model of OSI—The International Standards Organization (ISO) Reference Model of Open Systems Integration (OSI) highlights the communication protocols required in a communications system.
3. Elemental model—As part of the initial system design, control tasks must be assigned to major system elements. A commonly used model is to divide the physical resources of the communication system into three segments and assign control tasks to one or another of the segments in order to achieve the required system performance characteristics. This model has no recognized name. It will be referred to here as the Elemental Model of System Control.

The final system design should provide for rapid and effective management, straightforward end-to-end communications, and cost-effective design featuring a rational assignment of control functions to major system elements.

CSOC will function as an operational resource of the Air Force Satellite Control Network and will interface with other AFSCN elements and with the National Aeronautical and Space Administration Network. The relationship of CSOC to other elements in the network is shown schematically in Figure 5.7. Intrasegment interfaces will exist between elements of one segment, while intersegment interfaces will exist between segments of CSOC. Interfaces such as SPADOC, the Department of Defense (DoD), and the National Aeronautics and Space Administration (NASA), as well as collocated programs, are identified as external interfaces.

5.12.5   Intersegment Interfaces

The SOC and SOPC segments interface for the planning, preparation, and control of SSV payload launches, in-orbit operations, and recoveries. SOC/SOPC flight-control interfaces include the real-time interchange of payload commands and telemetry, acquisition and tracking data, and SSV telemetry, command, voice, and text data. Both SOC and SOPC interface with supporting systems for system configuration and status data.

SOC and SOPC interface through the CS for the exchange of data, just as the SS, VS, and NCS segments support SOC and SOPC through interfaces with the CS. The FS interfaces with the other CSOC segments to provide space, power, lighting, fire protection, and other facility-related support, including some fixed assets required to support physical security. The VS supports SOC and SOPC and other CSOC segments by providing training for their staffs and for flight and ground crews using a variety of simulated functions.

5.12.6 ISO Reference Model

The ISO Reference Model of Open Systems Integration (OSI), illustrated in Figure 5.15, is a general seven-layer framework for protocol development. The ISO Reference Model was originally developed for civil communication net-

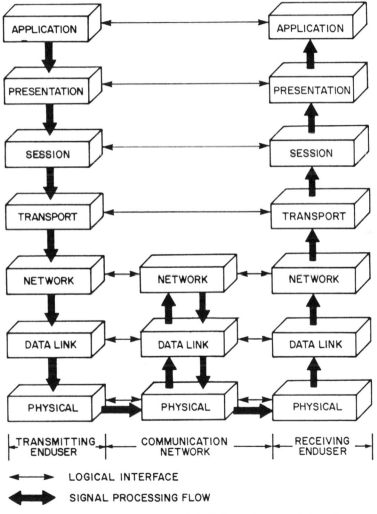

Figure 5.15 ISO Reference Model of Open Systems Integration.

works, but it is rapidly becoming a standard for all networks. (This model is discussed in more detail in Chapter 8.) As its name implies, it is intended to simplify the task of integrating different systems. For example, when integrating a local-area voice and data distribution system into a worldwide SATCOM system, the users, as well as the networks, may require new interface protocols and processing abilities. The Reference Model simplifies design of these new techniques, by subdividing the necessary tasks into several nearly self-contained classes. Each class, or protocol layer, can be designed in relative isolation from the other six. The Reference Model can also be applied to the design of a single autonomous system. Here as well, it subdivides the complete set of protocols and procedures into seven nearly self-contained categories.

Unlike the DCS SYSCON model, which focuses on overall network operations, the OSI Reference Model focuses on the individual user-to-user processing and control of communication channels. These protocols are embedded within the network's processors and within the end user's communication terminals and applications equipment. The relationship between DCS SYSCON and the ISO Reference Model is illustrated pictorially in Figure 5.16 (the figure also refers to the Elemental Model discussed in Chapter 4). DCS SYSCON is oriented toward control of common communication resources. The ISO Reference Model is oriented toward control of user-to-user communications. The two models overlap within the network, where users must conform to the protocols of the entire user community.

The Reference Model satisfies several objectives intended to simplify communication system design: (1) each layer communicates logically with an equivalent layer at the other end of a real or virtual communication link; (2) each layer provides a transparent circuit to the layer above it; (3) each layer represents a logically consistent set of processing functions; and (4) processing tasks are not split between levels. These objectives allow each layer to be designed in relative isolation from all other layers, simplifying control system development.

To illustrate these concepts, let us consider a voice telephone circuit connecting a European commander and the President of the United States. Their conversation, in human terms, is intended to achieve certain goals such as a transfer of information, commands, etc. These goals represent the application layer of the ISO model. This layer is defined by the end user's task. During the conversation, each user interacts with the network by voice. This protocol, which was selected by the users when they decided to use the telephone, is defined by the presentation layer. In like manner, information is successively processed by various equipment until it is transmitted over a communication circuit as an electromagnetic waveform. The modulation technique, and control overhead intended to maintain a point-to-point link, is part of the physical and link layer protocol.

In a communication network with many physical nodes, the nodes are connected by physical circuits operated in accordance with appropriate physical layer protocols. Each node also processes communication traffic in accordance with user requirements to the network, and maintains connectivity with the

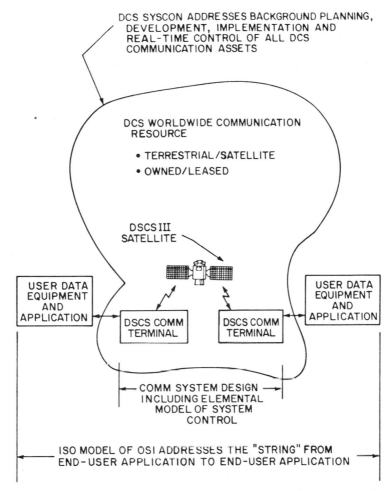

DCS SYSCON ADDRESSES BACKGROUND PLANNING, DEVELOPMENT, IMPLEMENTATION AND REAL-TIME CONTROL OF ALL DCS COMMUNICATION ASSETS

DCS WORLDWIDE COMMUNICATION RESOURCE

- TERRESTRIAL/SATELLITE
- OWNED/LEASED

DSCS III SATELLITE

USER DATA EQUIPMENT AND APPLICATION

DSCS COMM TERMINAL

DSCS COMM TERMINAL

USER DATA EQUIPMENT AND APPLICATION

COMM SYSTEM DESIGN INCLUDING ELEMENTAL MODEL OF SYSTEM CONTROL

ISO MODEL OF OSI ADDRESSES THE "STRING" FROM END-USER APPLICATION TO END-USER APPLICATION

Figure 5.16 Typical relationship among three models of system control.

network via network and data link protocols. The users control their end-to-end communications via higher-level protocols—transport and above.

As modern satellite communications systems strive for increasing scope, efficiency, worldwide connectivity, and interoperability with other systems, a standardized protocol model becomes necessary. As a key example, the MILSTAR will be a combined strategic/tactical, mobile SATCOM system offering multiple antijam (AJ) and low-probability-of-intercept (LPI) waveforms, worldwide connectivity, autonomous operation, satellite on-board signal processing, and demand-assigned communication and control channels for increased system efficiency.

The MILSTAR control system, currently under development, incorporates a top-down protocol hierarchy nearly identical to the ISO Reference Model. The similarities are highlighted in Table 5.5. Both hierarchies include several layers that perform similar functions.

The high-level layers, transport and above, are normally influenced by user requirements rather than networking considerations. Users, among themselves, determine (1) the types of messages they will employ, (2) how they will initiate and maintain communications with one another (as opposed to maintaining communications with the network), (3) how data will be presented and massaged, and (4) the use of that data. However, the trend toward enhanced interoperability will require increased consideration of these issues by the network designer. For example, the network may be required to offer a gateway service between dissimilar subnetworks or user classes. This requirement might be satisfied by the use of special-purpose processors contained within the network, designed to interface between dissimilar presentation, session, or transport protocols. While these processors would be physically contained within the network, they would not be network-level protocols. They would be classified according to the protocol layer they manipulate. Such structures are described in Section 5.12.5.

The MILSTAR network layer will support resource control, routing, flow control, and network access. These functions will be performed by a satellite on-board processor called the Resource Controller (RC). This computer will perform all real-time management and control of the SATCOM system, making it independent of nonsurvivable ground-based control centers. The RC will allocate control and data channels in response to prioritized user requests, calculate routing through the cross-linked satellite network, respond dynamically to system failures, continually search for, and provide service to, new authenticated users, and perform real-time reconfiguration in response to changing user requirements and environmental conditions. Network access is differentiated from session management in the following way: network access authenticates a user's right to access the network, in this case, the RC on-board the satellite; session-layer protocols authenticate users to each other. When perceived from the ISO Reference Model, the RC implements network-level protocols. When perceived from DCS SYSCON, the RC supports and directs all real-time system control functions associated with network control, traffic and routing control, performance monitoring, and technical control.

The MILSTAR data link layer supports signal processing and frame formatting. The satellite must read user control messages and process user data streams down to the information bit level. User control messages must be read on-board the satellite in order to support autonomous network control by the RC. These protocols, therefore, provide service to the network-layer protocols one level up. User data streams are processed for several reasons. Processing of data down to the information bit level provides significant AJ, since uplink noise and jamming processes are dissociated from power amplification from the downlink. Signal processing also allows extremely flexible satellite-based routing, multiplexing, data combining and frame formatting. These capabilities allow uplink and downlink modulation formats and framing to be optimized separately for each communication regime and user class.

Table 5.5 Functional description of protocol hierarchy.

Protocol Layer	Representative Functions in ISO Model	Potential Functions in MILSTAR
Application	End User	End User
Presentation	Data reformatting; encryption, data compression, display formatting	Data reformatting; voice/digital conversion
Session	Establish connection, sign on, authentication, validation, real-time management	Network activation; real-time management
Transport	End-to-end delivery, multiplexing, message assembly; end-to-end error control	End-to-end data delivery; error correction
Network*	Routing; congestion control, service accounting, and diagnostics	Resource control; routing; flow control; network access
Data link	Frame formatting, node-to-node error control and flow control	Signal processing; frame formatting
Physical	Electrical and mechanical interface; procedural control; functional meaning of signals	Transmission/reception of signals

*This layer is identified as ''Access Control'' within MILSTAR.

The MILSTAR physical labor protocols describe modulation techniques used for transmission and reception of signals. These techniques provide AJ and LPI capability and timing synchronization.

5.12.7 Elemental Model of System Control

The Elemental Model of System Control, illustrated in Figure 5.17, is chiefly applicable to the design of satellite communication systems. It highlights two major technological tradeoffs involved in control system development: (1) whether the control architecture should be centralized or distributed and (2) which control tasks should be supported at each control node. The Elemental Model consists of two major phases. The first phase categorizes major control functions into three areas: (1) those necessarily performed at the ground communication terminals, (2) those necessarily performed on board the satellite, and (3) those required for network control. This process is related to user requirements and technological state-of-the-art. An example of this process is given in Table 5.6 for the DSCS. The second phase determines support techniques for the network control functions. These functions can be performed at one or several dedicated ground control stations, as in the DSCS III RTACS, on board the satellite as in MILSTAR, at the individual satellite communication terminals in a distributed architecture, or some combination of techniques. The chief design issue is selection of a network control architecture that meets all user requirements in a cost-effective manner. The major design choices are between distributed and centralized control, the location of the control node(s), and functions performed at (each) control node.

The Elemental Model represents a top-down approach to system design. This choice becomes increasingly important as SATCOM systems strive to support ever-wider user communities. In MILSTAR, the design of the space segment and the terminal segment is divided among the defense services (primarily Navy and Air Force). With early assignment of control functions to major elements of the SATCOM system, the Elemental Model simplifies interservice coordination during design and integration of system elements.

A critical feature of this design technique is that it progresses in parallel with communication system design. This approach generally results in several design iterations before a final system architecture successfully integrates all control and communication requirements into a single feasible and cost-effective design. The problem here is that control system design must be based on assumptions about the available communication hardware and communication system design must be based on assumptions about control system protocols and requirements.

The MILSTAR system illustrates the interplay between control system design and communication system design. Current planning calls for multiple satellite spot beams to cover the entire earth's surface visible from the satellite. The control system must be designed to cope with partial coverage and part-time coverage of potential and ongoing users. Examples of control functions affected by this constraint are (1) dissemination of system time, (2) dissemination of

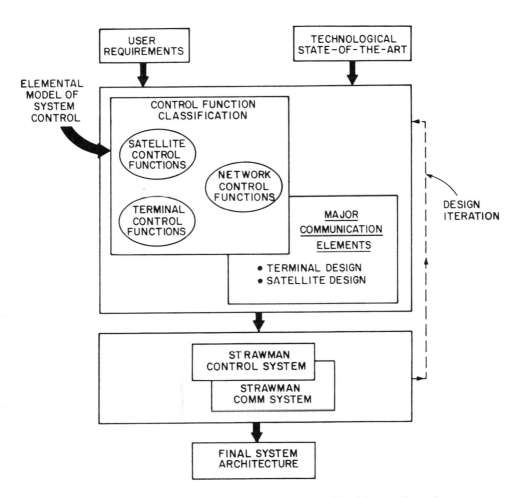

Figure 5.17 Design environment for Elemental Model of System Control.

**Table 5.6 Classification of DSCS control functions within the
Elemental Model of System Control [6].**

Satellite Control Functions

- Control and monitor the payload
- Respond to directives from network control element(s)
- Obtain environmental measurements
- Provide performance and environmental reports as required to network control element(s)
- Control satellite housekeeping functions

Terminal Control Functions

- Control and monitor terminal equipment
- Monitor and report link quality
- Respond to all directives from the network control element(s)
- Maintain transmission parameters within prescribed allocations
- Maintain calibration of terminal equipment
- Schedule and control maintenance actions
- Obtain environmental measurements
- Downlink timing acquisition
- Uplink ranging control

Network Control Functions

- Allocation of system resources
- Adaptive control and system reconfiguration
- Timing control
- Performance measurement and reporting
- Failure detection, isolation, reporting and analysis
- Monitoring and control of net terminal performance
- Monitoring and control of orderwire channels
- Coordination of operational data with terminals

satellite ephemerides, (3) loop-back ranging between earth terminals and the satellite, (4) polling for service, and (5) searching for users attempting to enter the network. These functions must be designed to cope with limitations imposed by the communication system (which is designed to meet the perceived jamming and detection threats). In the other direction, the requirement to perform control tasks without disrupting communications places a constraint on the communication system design. In particular, FDMA and TDMA framing protocols, typically part of the communication system, must be designed to allow transmission and reception of control messages without interrupting or dropping communication channels.

Throughout the design process, changes to the control system may be forced as a result of an increasing understanding of the control problem, as well as a changing concept of the baseline communications configuration.

Focusing on the Elemental Model itself, Table 5.6 illustrates, relative to DSCS, major control tasks that must be allocated to the space segment, the terminal segment, and network control node(s). Final system survivability, responsiveness, and cost are critically dependent on the way network control functions are supported. The DSCS III control employs centralized network control facilities. These centralized ground stations interface with the DCS hierarchy at the theater and worldwide levels. MILSTAR, representing state-of-the-art technology and stringent constraints on survivability and autonomy, will perform all real-time network control functions with satellite on-board resource controllers. But ground control stations will still exist to perform background functions, maintain management control, and provide control backup in the event of resource controller failure.

The interaction between system models is frequently subtle. As an example, consider the interaction between the ISO Reference Model and the Elemental Model within DSCS and MILSTAR. A major function of the NCF is real-time power budgeting and link control within the user community. This function is extremely complex because of the basic FDMA scheme used in DSCS, combined with enormous variations in user power levels, data rates, quality requirements, and environment (including jamming). Each NCF controlling operations for a single DSCS III satellite continually checks and optimizes user transmit power levels in order to maximize the utility of the SATCOM system. This operation represents a significant computing load for the SATCOM control system, which the Elemental Model perceives as a functional control requirement. The requirement must be assigned to a control node in such a way that technological constraints and other user requirements are satisfied. In the DSCS control, this function is performed at centralized ground-control stations (NCF). In essence, the control architecture was influenced, to some extent, by earlier decisions on physical-layer protocols.

For MILSTAR, the control architecture and hierarchical protocols are being developed simultaneously. The result is a well-integrated design that balances user requirements and technological limitations. The critical system requirements of AJ, LPI, survivability, autonomy, worldwide interoperability, and connectivity have lead to the development of a signal-processing satellite with on-board real-time control. Protocols for the physical, data link, and network layers of the ISO Reference Model were influenced by the need to support signal processing and information processing (of control streams) on board the satellite.

5.13   FUTURE SATELLITE TRENDS[4]

Several trends in satellite communications promise enormously more capable SATCOM systems. These trends are supported by rapid advances in electronics and networking theory and plummeting costs of space and terrestrial hardware (for a given capability). Future SATCOM systems are likely to exhibit increasingly intelligent satellite-based controllers, highly proliferated satellite constellations for enhanced physical survivability, and distributed/redundant networking architectures based on intelligent terminal control elements and extensive interoperability among all elements of the DCS community (see Figure 5.18).

Intelligent satellite-based controllers are appearing for the first time with MIL-STAR. The future will see these controllers drop in price and become increasingly adaptive to changing environmental conditions. Increasingly, background control functions such as performance monitoring and assessment, and reconfiguration control, will be performed on the satellite. Viewed from DCS SYSCON, many of the real-time control functions performed at the worldwide and theater levels will be automated on the satellite. Under nominal conditions, DCAOC and the ACOCs will perform only background control and management review. However, a backup control capability will be maintained to guard against satellite equipment failures. In terms of the ISO Reference Model, an increasing number of network-level control interactions will terminate on the satellite. This trend will be balanced by the introduction of space-to-ground control links carrying summary information and data base updates in order to maintain current data bases and review status at the ground-based backup control center(s).

As space hardware becomes less expensive, the number of spacecraft per system will proliferate. This trend is already apparent in such systems as MIL-STAR and GPS (which, under current planning, will have 18 spacecraft in the active constellation with 6 on-orbit spares). Proliferated systems are inherently robust against conventional physical and jamming attacks.[5] This concept, known generically as ''cloudstat,'' has a large number or ''cloud'' of spacecraft, interconnected with crosslinks, that provide communication support and worldwide

---

[4]This section represents the authors' viewpoint. The technological projections described here do not represent official planning for the DCS.

[5]An enemy must destroy, or effectively jam, all satellites in view of a particular user in order to deny communications to that user. This task becomes increasingly difficult as the number of satellites increases, and as individual satellites become more resistant to jamming. However, if the SATCOM system is required to survive the effects of high-altitude nuclear bursts, additional design and cost issues must be resolved.

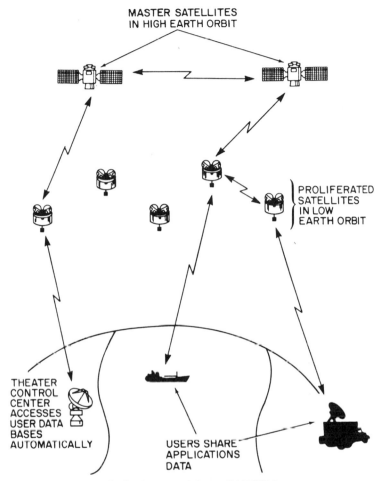

MASTER SATELLITES
IN HIGH EARTH ORBIT

PROLIFERATED
SATELLITES
IN LOW
EARTH ORBIT

THEATER
CONTROL
CENTER
ACCESSES
USER DATA
BASES
AUTOMATICALLY

USERS SHARE
APPLICATIONS
DATA

Figure 5.18 Integrated future SATCOM system.

connectivity. The system could be implemented with a number of identical, equally capable satellites. Alternatively, the system could include a large number of relatively inexpensive satellites with a smaller number of more capable, "master" spacecraft (see Figure 5.18). In terms of DCS SYSCON, as well as the Elemental Model, this latter concept raises questions of overall coordination, orbital determination, and distribution of system time. The traditional approach would be to perform these functions on the ground, with control traceable to the theater and worldwide levels of the DCS hierarchy. However, future systems may distribute these functions to the space segment (some or all of the proliferated constellation) or to a large number of earth terminals (earth terminal enhancements are discussed below).

As computer control equipment becomes more reliable, capable, and afford-able, individual SATCOM terminals will become more intelligent (regardless of size and mobility constraints). In the distant future, it may be possible to support networkwide control functions out of any user terminal. This capability could lead to massive proliferation of critical ground control nodes, making the overall system truly immune to single-point and multi-point failures (the individual terminals could be organized in a precedence list, so that every potential control terminal has a designated backup). Intelligent terminals could assist in identifying network faults, as well as offer enhanced user services such as interoperability with terrestrial communication networks (enhanced intelligence would support a reconfiguration gateway function as standard equipment in most or all SAT-COM terminals). Currently, such distributed architectures face major disadvan-tages in terms of control system overhead, cost of individual components, and reliability of fail-safe and robust control protocols. These problems can be viewed in terms of the Elemental Model (for cost and capability of individual compo-nents) and the ISO Reference Model (for control overhead and development of necessary control protocols).

Finally, work by the Navy [7] points the way toward enhanced system inte-gration that would allow widely distributed control centers and users to access data bases and support functions throughout the worldwide DCS. At present, internetwork and intersystem interfaces require careful design and implemen-tation on a case-by-case basis. If general-purpose gateway functions could be supported by the communication network, rather than the users, such interfaces would be vastly simplified (at least from the users' perspective). Personnel at a theater-level ACOC could, for example, access user data bases at any facility connected to the worldwide DCS. Alternatively, users could share data bases for such time-critical functions as target acquisition and pointing. This capability would be supported by enhanced routing and accessing capabilities throughout the satellite-based assets of the DCS, as well as gateway protocols supported at most or all communication terminals (satellite and nonsatellite).

These gateway functions would support protocol translation that would allow incompatible users and equipment to share data. From a user perspective, this is primarily a question of protocol translation perceived within the ISO Reference Model. From a worldwide control perspective, this trend would integrate aspects of DSCS SYSCON with the ISO Reference Model. For example, control and coordination channels within DCS SYSCON are generally viewed as commu-nication channels by the system designer, since the data flowing over these channels do not directly control the communication assets supporting the flow. However, with automatic gateway support and computer-to-computer commu-nication across system boundaries and over large distances, the higher-level functions in the ISO Reference Model must be supported within the traditional boundaries of the communication system in general, and DCS SYSCON in particular.

To summarize, future control of military satellite communications will tend to integrate major facets of the three control models in use today. Many functions traditionally performed at the sector and nodal levels will be distributed up to the theater and worldwide levels (or the satellite) and down to the station level (where intelligent terminals will assist in overall system control). Traditionally user-oriented levels of the ISO Reference Model will be supported by the communication system itself. This approach will lead to highly efficient data base transfers and data sharing among users and control centers. As satellite constellations proliferate and individual satellites become more capable, real-time control functions will be performed on board the satellite(s) and distributed networking protocols will become increasingly important.

5.14 REFERENCES

1. Spilker, J. *Digital Communications By Satellite*. Englewood Cliffs, N.J.: Prentice-Hall, 1977.
2. Bond, F. "Future Trends in Commercial and Military Systems." *Conference Record*, Vol. 2, ICC, 1983.
3. Rosner, R.D. "An Integrated Distributed Control Structure for Global Communications." *IEEE Transactions on Communications* (September 1980).
4. *POLICIES: System Control of the Defense Communications System*. DCA Circular 300-50-6, Defense Communications Agency, 24 June 1977.
5. Ulsamer, E. "The Growing, Changing Role of C^3I." *Air Force Magazine* (July 1979): 36–48.
6. Eisenberg, R.L. "JTIDS System Overview." *Principles Operational Aspects Precision Position Determination System*, NATO Advisory Group Aerospace Research and Development AGARDograph 245 (July 1979): 26-1–26-7.
7. Allgaier, G.R. "Command Center Network (CCN)—Backbone of Future Command and Control." *Naval Engineers Journal* (December 1980).

5.14.1 Further References

1. Beaumont, R.A. "The Tactical Spectrum and C^3 State Variance: Accommodating Uncertainty." *Signal* (March 1981): 45–49.
2. Brandon, W.T. "Design Tradeoffs in Antijam Military Satellite Communications." *IEEE Communications Magazine* (July 1982).
3. Bush, H. "Multifunction Communications and Tactical Data Links." *Techniques for Data Handling in Tactical Systems II* (AGARD Conference Proceedings 251), Monterey, California, October 1978.
4. Cherry, C. *On Human Communications; A Review, A Survey, and A Criticism*, 3rd ed. Cambridge: MIT Press, 1978.
5. Conner, W.J. "The AN/TRC-170—A New Digital Troposcatter Communication System." *Proceedings of Techniques Seminar, Digital Microwave Transmission System*, Department of Electrical Engineering, Princeton University, Princeton, N.J., pp. 19–23, 27 February 1979.
6. Feldman, N.E., et al. *Writer-Reader Delays in Military Communications Systems*. The Rand Corp., R-2473-AF, October 1979.
7. Gardella, R.S. *Issues in the Design and Use of Secure Terminals*. The Mitre Corp., Mitre Technical Report 3128, June 1976.

8. Gleason, M.R. and LaBanca, D.L. "A System Overview of Tactical Satellite Communications." *Proceedings of New Techniques Seminar, Digital Microwave Transmission System*, Department of Electrical Engineering, Princeton University, Princeton, N.J., pp. 30–48, 27 February 1979.

9. *IEEE Proceedings on Military Communications*, MILCOM, 1983.

10. Keegan, J. *The Face of Battle*. New York: Random House, 1977.

11. Kirchhofer, K.H. "Key Management in Cryptographic Systems." *International Defense Review* (September 1980): 1396–1398.

12. LaVean, G.E. "Interoperability in Defense Communications." *IEEE Transactions on Communications* (September 1980).

13. McKee, W.P., Jr. "Architectural Considerations for the Defense Communications System of 2004." *IEEE Communications Magazine* (July 1982).

14. Ricci, F.J., Heppe, S.B., and Arozullah, M. "Control Concept for DAMA Communications in a Future DSCS." STI/E-TR-25025, 23 July 1982.

15. Schutzer, D., Grace, J., Halpeny, O., Ricci, F. "DCS System Control Concept Formulation." Defense Communications Engineering Center, Technical Report No. 5-74, February 1974.

16. Tanenbaum, A.S. *Computer Networks*. Englewood Cliffs, N.J.: Prentice-Hall, 1981.

17. Thompson, T.H. "Tactical Air Forces Integrated Information System Master Plan." *Signal* (August 1978): 68–73.

18. "TRI-TAC." Joint Tactical Communications Office (TRI-TAC Office), Tinton Falls, New Jersey, September 1979.

19. Tyree, B.E. et al. "Ground Mobile Forces Tactical Satellite SHF Ground Terminals." *Proceedings of New Techniques Seminar, Digital Microwave Transmission System*, Department of Electrical Engineering, Princeton University, Princeton, N.J., pp. 49–53, 27 February 1979.

Chapter 6

UNIFIED COMMUNICATIONS INFORMATION FRAMEWORK

In the last decade, we have experienced some trends that may serve to dramatically alter warfare in rather fundamental ways. It is believed that to realize the full benefit of these trends and changes, the various information-manipulative functions of communications, surveillance and reconnaissance, intelligence, cover and deception, electronic warfare, communications and operations security, and command and control need to be viewed, designed, and managed from a single unified framework. This framework should provide for the required force multiplier effect we have described earlier. Communications systems alone cannot ensure effective information flow and in many cases they may represent a potential Achilles heel. The larger picture, a unified network, demands consideration.

Today we are on the threshold of major new innovations and trends that will result in order-of-magnitude increases in both the tempo (rate of change or action) and range lethality (action-at-a-distance) normally associated with warfare. Satellites that are capable of communicating sensed intelligence, literally at the speed of light, hover in the sky. C^2 (Command and Control) platforms in space and weapons that can be launched from space are becoming technically viable. Missiles and torpedo technologies have progressed to the point where unmanned vehicles can be launched at intercontinental ranges and very high speeds. They can fly ballistic profiles, they can skim the earth's surface to avoid detection, and they can execute pop-up terminal maneuvers. They possess sophisticated onboard sensors and processors providing a significant degree of autonomy. Sophisticated remotely piloted vehicles (RPVs) that extend a platform's sensor range against targets "over-the-horizon" are also becoming technically viable.

Current concepts of space and time will be dramatically altered with the introduction of these new innovations and trends. And, although the basic warfare principles such as "concentration of forces" and "surprise" still are valid objectives, the means and mechanisms to best achieve these objectives are changing.

Consider the notion of "time-on-target," which is the execution of a strike by units dispersed over an ocean or a continent that is timed and coordinated to produce the arrival of their weapons at targeted sites in near-simultaneous waves.

An unprecedented degree of precise timing and coordination is required to produce this effect. Further, the initiation of such a strike may be based on a response to a chain of events that precede the decision to attack. It may not be possible to forecast such a chain of events or to anticipate too far in advance. And, since the attack could well involve ships, missiles, and aircraft far from the area of interest, our standards concerning what constitutes an acceptable state of "combat readiness" may also have to change substantially.

On the other hand, the effectiveness of our defenses can be expected to improve manifold. The capability of our forces for self-protection can also be expected to increase dramatically.

All trends indicate significant growth in the size, armor, and defensive capabilities of our forces. These forces will literally become fortresses. They already possess an arsenal of sophisticated defensive-warning radars, electronic support measures (ESM), antimissile missiles, electronic warfare (EW), and electro-optic (E-O) countermeasures, and offboard decoys to defend themselves against incoming attack. In the future, we can expect still greater capability in these areas of defense as well as the employment of such new technologies as laser and particle beams. Future enemy raids will have to traverse layers of defensive zones or barriers before they can reach their targets.

Clearly, the offensive must be prepared to make the most of the element of surprise and concealment if an attack is to succeed. Even so, the offense must be prepared to suffer unprecedentedly high attrition rates. When one lives in an environment of continual worldwide surveillance, achievement of the element of surprise takes on a new dimension. Techniques for the concealment from, and the coordinated deception of, the enemy's surveillance systems and warning indicators need to be developed, perfected, and exercised.

All of these considerations highlight the need to assess, evaluate, coordinate, manage, and control one's forces and to possess timely and accurate data concerning the enemy force's deployments, capabilities, characteristics, modus operandi, and intentions. Future weapons will be "smart" enough to travel over great distances at supersonic speeds, to sense and maneuver to avoid detection and obstacles or weapons placed in their path, and to successfully pick out, jam, and/or engage a target in a heavily cluttered environment despite enemy deception measures. This capability requires that these weapons be supplied with a great deal of precise and timely intelligence concerning the tactical situation, including enemy capabilities, signatures, and characteristics, in order to successfully penetrate deep into heavily protected enemy defense zones.

These future capabilities will give the commander great flexibility in adapting to a situation in innovative and creative ways. He will be harder to stereotype, more capable of surprise, and thus, harder to defeat. On the other hand, his enemy will likely be equally sophisticated. The commander will be faced with an extremely complex environment and a bewildering array of options and decisions to be made, and because of the greatly increased "tempo" of operations, he will have little time to evaluate a situation and to select and effect a course of action.

Such a situation requires organization and integration of the various information and signal collection, dissemination, management, and control disciplines and processing techniques into a cooperative team of "expert" modules or assistants that can work together effectively to accomplish the objectives of "putting metal on the target," orchestrating a situation, and making sound decisions under tight time constraints. This level of C^3 necessitates a unified framework from which to design, manage, and coordinate the various information-manipulation functions.

6.1 FRAMEWORK

Consider the various information-manipulative functions of command, control, and communications, surveillance, reconnaissance and intelligence, cover and deception, and electronic security. Each of these functions has a role in direct support of war fighting. Some of them have the function of maximizing the information available to hostile forces [1].

For the cases of information exchange (e.g., communications) and active (or noncooperative) search (e.g., radar), the increase of knowledge that results is related to the amount of knowledge of the situation already possessed. The state of knowledge should increase in direct proportion to one's own ignorance of the current situation. That is, the more there is to know, the more there is to gain by actively searching, collecting, and exchanging information about the situation among the various concerned and cooperating parties. This relationship is expressed by the following equation:

- Case 1: Exchange of information and active search

$$\dot{u} = -au$$

where (6.1)

u = entropy, or ignorance of friendly

$\dot{u} = \dfrac{du}{dt}$ = rate of change in entropy

a = positive constant.

Information may also be collected and exchanged by means that exploit such actions and activities as enemy communications, radar emissions, or patterns of behavior and that are dependent for success on the enemy's ignorance concerning the nature, location, means, or method of the collector and receiver (e.g., the interception, geolocation, and decoding of enemy communications and radar transmissions). For these cases, the increase gained in one's knowledge of the situation is proportional to both one's current knowledge and the enemy's current ignorance of the situation. The more knowledgeable we are, and the more ignorant the enemy is, of the current situation, the more effective we will be in our covert collection attempts. This relationship is expressed by the following equation:

● Case 2: Covert and cooperative collection

$$\dot{u} = -b(1-u)\, v$$

where

u = entropy of friendly forces   (6.2)

v = entropy of enemy forces

$\dot{u}$ = rate of change in entropy of friendly forces

b = positive constant.

Cover and deception and electronic warfare are concerned with interfering with and inhibiting the enemy's information collection and exchange activities with the goal of deceiving, confusing, and degrading the enemy's knowledge of the situation (e.g., by use of jamming, delay, and decoys). For these cases the success we achieve in degrading the enemy's knowledge seems to be directly related both to our own and the enemy's current knowledge of the situation. The more we know, the more effectively we can manipulate and inhibit the enemy's attempts at information collection and exchange. And, the more the enemy knows, the more he has to lose through our use of cover and deception and electronic warfare. For example, if the enemy knew very little about the current situation, he would not lose much if we were to attempt to jam or spoof his communications circuits. This relationship is expressed by the following equation:

● Case 3: Active jamming and deception

$$\dot{v} = c(1-u)(1-v)$$

where

u = entropy or ignorance of friendly forces   (6.3)

v = entropy or ignorance of enemy forces

$\dot{v}$ = rate of change in entropy of enemy forces

c = positive constant.

Finally, it should be noted that even for the case when no information-manipulation activity is occurring, the information states do not remain static. The situation is continually changing and without an active effort at collecting and exchanging information concerning the evolving situation, our state of knowledge will degrade and our ignorance will grow in direct proportion to our past state of knowledge. This relationship is expressed by the following equation:

- Case 4: Normal growth in uncertainty

    $$\dot{u} = d(1 - u)$$

    where   (6.4)

    u = entropy or ignorance of friendly forces

    $\dot{u}$ = rate of change in entropy of enemy forces

    d = positive constant.

Various operational and design measures can be taken to minimize an enemy's effectiveness and to improve one's own capability in performing these information-manipulative functions. For example, security measures serve to make the enemy's job of covert intelligence collection more difficult by denying him key sources of data. These measures can be achieved through the application of both appropriate operational and design measures. Fixed operational patterns are avoided, with secrecy and maximum security characterizing the planning and execution of all operational plans and strategies. Likewise, communications and active surveillance systems are designed to have low-probability-of-intercept (LPI) with minimum detectable energy over any given frequency band or geographic location. Generally, these objectives are achieved through such varied techniques as power control, spread spectrum, frequency hopping, narrow antenna beams, and bistatic operation.

Communications signals are scrambled (encrypted) to deny the enemy the capability to extract any intelligence from the signal content. The use and scheduling of communications and active surveillance emissions are staggered across time, space, and frequency in an unpredictable fashion so as to appear random to the enemy. The result of these measures can be reflected by an appropriate adjustment of the constant coefficient b of (6.2) downward; b becomes a smaller number, thus requiring a greater exertion of effort on the collector's part to increase his knowledge of the situation. Likewise, communications and active surveillance systems can be designed to be more robust and resistant to enemy jamming and deception countermeasures. These objectives are accomplished through such design techniques as frequency agility, spread spectrum, redundancy, high adaptive antenna gain and selective nulling, monopulse, selective polarization, and adaptive routing and rerouting. Generally, these techniques involve a tradeoff between the available channel capacity (information exchange rate), the acceptable error rate, and the degree of jam resistance. These measures result in an adjustment of coefficient c of (6.3) downward.

Although these various information-manipulative functions all have the common objective of increasing the friendly force's knowledge of the situation while degrading and spoofing the enemy with respect to his knowledge of the situation, when viewed in isolation from one another they could well be employed at cross purposes and end up substantially negating one another. For example, exercising emissions control prohibits the employment of active search, and jamming an enemy's communications prevents covert collection and exploitation of these communications.

The respective designs of the systems that support these various information-manipulative functions should be implemented and managed in a coordinated fashion so that the systems would be capable of working together to the maximum extent possible. For example, EW, surveillance, and communications systems should be designed so that simultaneous employment of these assets will not cause them to jam or interfere with one another.

When viewed in this manner, the strategy illustrated in Figure 6.1 emerges as a reasonable information-handling systems design and control scheme. The objective of this control strategy is to maximize a measure of the combined state of knowledge of one's own forces and state of ignorance of one's opponent's forces through judicious application of information warfare. Friendly sensor

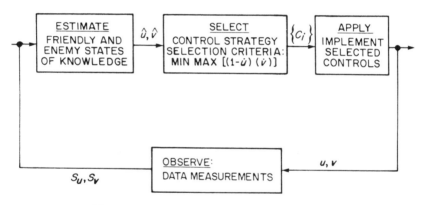

Figure 6.1 Information warfare control strategy.

measurements, intercepts of enemy communications and sensor radiations, and human intelligence reports are received, combined, and interpreted to form two pictures: one representing friendly's state of knowledge of the situation and the other representing the opponent's state of knowledge of the situation. The first picture is friendly's best assessment of the current tactical situation (i.e., what units are located where, their course, and likely destination and intentions) with indications of the confidence associated with these estimates. The other picture is friendly's best estimate of his opponent's assessment of the situation and associated uncertainties. These pictures are used to estimate both friendly's and enemy's status of knowledge (and ignorance) of the current situation.

The various design concepts and techniques briefly discussed here are employed to improve one's capability and effectiveness in information warfare. The details of these design concepts and techniques with respect to their utility, performance, and implementation are described and explained in some depth in later sections of this book.

6.2   ESTIMATION OF THE STATES OF KNOWLEDGE AND IGNORANCE

The estimate of the state of ignorance would be proportional to the sum of uncertainties associated with a unit's identity, location, and course or destination. Misinformation would correspond to a case where uncertainty is low, but where an error bias has been introduced causing a serious misidentification or mislocation. This situation will result in a larger entropy measurement than for the state of total ignorance.

As an illustration of this point, consider the example in Figure 6.2. In this example there are ten location cells and two ships (course and destination information have been omitted for simplicity). Each location cell has three terms, namely:

$$- q_j (A) \ln P_j (A/A) - q_j (B) \ln P_j (B/B) - q_j (\overline{A}, \overline{B}) \ln P_j (\overline{A}, \overline{B}/\overline{A}, \overline{B}).$$

Where $q_j(A)$ = probability that ship A is in location cell j;
$q_j(B)$ = probability that ship B is in location cell j;
$q_j(\overline{A}, \overline{B})$ = probability that neither ship B nor ship A is in location cell j.

A and B are mutually exclusive (ships A and B cannot be in same location cell at the same time):

$P_j (A/A)$ =   probability that, given ship A is at location cell j, ship A is detected;

$P_j (B/B)$ =   probability that, given ship B is at location cell j, ship B is detected;

$P_j (\overline{A}, \overline{B}/\overline{A}, \overline{B})$ =   probability that, given neither ship A nor B is at location cell j, no ship is detected.

If, as shown in Figure 6.2, ship A is actually at location cell (1, 1), or row 1 and column 1, and ship B is at location cell (10, 9), row 10 and column 9, then all the q terms are zero with the exception of

$$q_{(1, 1)} (A) = 1, q_{(10, 9)} (B) = 1 \text{ and } q_{(n, m)} (\overline{A}, \overline{B}) = 1.$$

for all n and m except for ($n = 1, m = 1$) and ($n = 10, m = 9$). Therefore, for Figure 6.2, the entropy reduces to the sum of 100 terms:

$$- \ln P_{(1, 1)} (A/A) - \ln P_{(10, 9)} (B/B) - \sum_{\substack{n, m = \\ 1,1 \text{ and} \\ 10,9}} \ln P_{(n, m)} (\overline{A}, \overline{B}/\overline{A}, \overline{B}). \quad (6.5)$$

For the case of total knowledge:

$$P_{(1, 1)} (A/A) = P_{(10, 9)} (B/B) = P_{(n, m)} (\overline{A}, \overline{B}/\overline{A}, \overline{B}) = 1 \quad (6.6)$$

and the entropy $= -100 \ln 3$.

For the case of total ignorance:

$$P_{(n, m)} (A/A) = P_{(n, m)} (B/B) = P_{(n, m)} (\overline{A}, \overline{B}/\overline{A}, \overline{B}) = \frac{1}{3} \qquad (6.7)$$

and entropy $= 100 \ln 3$.

For the case of complete misinformation:

$$
\begin{aligned}
P_{(1, 1)} (A/A) &= P_{(10, 9)} (B/B) \\
&= 0, \text{ as well as selected } P_{(n, m)} (\overline{A}, \overline{B}/\overline{A}, \overline{B}) \text{ terms, and entropy} \\
&= -n \ln 0 \\
\text{entropy} &= \infty.
\end{aligned}
$$

$$(6.8)$$

	1	2	3	4	5	6	7	8	9	10
1	A									
2										
3										
4										
5										
6										
7										
8										
9										
10								B		

Figure 6.2   Estimation of entropy.

6.3   INFORMATION WARFARE CONTROL STRATEGY

Information warfare is a two-sided contest. It is assumed that both parties will seek to select that combination of information-manipulation functions that serves their best interests. Accordingly, friendly should select that set of information-manipulative functions that results in a maximum objective function, assuming that the opponent will always select a corresponding set of information-manipulation functions designed to minimize the same objective function. This assumption corresponds to a worst case approach. It might prove desirable to provide the decision maker with other possible control strategies and their possible outcomes, such as one based on a best case analysis and another based on a most likely (derived from historical precedence) enemy response.

Of course, the end objective is to prevail in battle, maximizing the enemy's losses and minimizing one's own losses. It will be shown in a later section that the amount of knowledge, or information, a combatant possesses directly influences his combat effectiveness. Accordingly, a suggested objective function is $(1 - \dot{u})\dot{v}$, the product of one's own force's increase in "state of knowledge" and the enemy's increase in his "state of ignorance." This objective function increases when friendly's knowledge of the situation is increased and/or the opponent's ignorance of the situation is increased. It is recommended that prior to the selection for execution of any set of information-manipulation functions, a projection of battle outcome be determined that includes the likelihood of enemy initiation of conflict and the predicted losses and associated risks. This battle outcome prediction should be presented along with recommended information-manipulation control strategies so that a course of action may be determined that includes consideration of the initiation of conflict as well as pure information warfare actions.

6.4 CONTROL STRATEGY EXAMPLES

The control strategy just enumerated may be illustrated by the following two examples:

First Example: Suppose one wished to choose between two candidate information-manipulation control actions (active surveillance and covert collection) in response to an opponent who is currently practicing covert collection. The problem is formulated as a choice between the following two cases:

- Case 1: Friendly, active surveillance. Enemy, covert collection.

$$\text{Friendly entropy } \dot{u} = d(1-u) - au \qquad (6.9)$$

$$\text{Enemy entropy } \dot{v} = d(1-v) - b(1-v)u \qquad (6.10)$$

- Case 1: Objective function $= [(1-d) + (d+a)u]\,(d-bu)\,(1-v) \qquad (6.11)$

- Case 2: Friendly, covert collection. Enemy, covert collection.

$$\text{Friendly entropy } \dot{u} = d(1-u) - b(1-u)v \qquad (6.12)$$

$$\text{Enemy entropy } \dot{v} = d(1-v) - b(1-v)u \qquad (6.13)$$

- Case 2: Objective function

$$= [(1-d) + bv + du - buv]\,(d-bu)\,(1-v) \qquad (6.14)$$

Case 1, active surveillance, should be selected over case 2, covert collection, when

$$\frac{a}{b}\frac{u}{1-u} > v \text{ or } \frac{a}{b}\frac{u}{v} > (1-u). \qquad (6.15)$$

Active surveillance should be chosen when one's own state of knowledge is less than an adjusted ratio of one's own ignorance to opponent ignorance.

Second Example: A second example concerns the selection of covert collection versus friendly jamming. This problem is formulated below:

- Case 1: Friendly, covert collection. Enemy, jamming.

$$\dot{u} = d(1-u) - b(1-u)v + c(1-u)(1-v) \qquad (6.16)$$

$$\dot{v} = d(1-v) \qquad (6.17)$$

- Case 1: Objective function =

$$[1 - d(1-u) + b(1-u)v - c(1-u)(1-v)]d(1-v) \qquad (6.18)$$

- Case 2: Friendly, jams. Enemy, does nothing.

$$\dot{u} = d(1-u) \qquad (6.19)$$

$$\dot{v} = d(1-v) + c(1-u)(1-v) \qquad (6.20)$$

- Case 2: Objective function $= [1 - d(1-u)][d(1-v) + c(1-u)(1-v)]$

$$(6.21)$$

Case 1 should be selected when

$$u \leqq \left(1 + \frac{b}{c}\right)v - \frac{1}{d} . \qquad (6.22)$$

Friendly should jam rather than perform covert collection when his ignorance exceeds an adjusted measure of the opponent's ignorance.

6.5 KNOWLEDGE AND COMBAT EFFECTIVENESS

The amount of knowledge, or information, possessed by a combatant influences his combat effectiveness. This influence can be represented as an adjustment of parameters in the differential equations that govern the combat outcome with respect to the number of units involved. These equations, which are discussed in great depth in the Lanchester combat-theory literature, are summarized in Figure 6.3, where x = number of friendly units and y = number of enemy units [2].

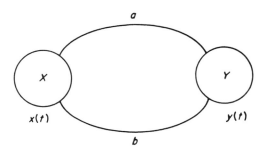

Attrition Process	Differential Equations	State Equation
Aimed fire vs. aimed fire	$\dfrac{dx}{dt} = -ay$ $\dfrac{dy}{dt} = -bx$	Lanchester (1914) $b(x_o^2 - x^2) = a(y_o^2 - y^2)$ Square Law
Area fire vs. area fire	$\dfrac{dx}{dt} = -axy$ $\dfrac{dy}{dt} = -bxy$	Lanchester (1914) $b(x_o - x) = a(y_o - y)$ Linear Law
Aimed fire vs. area fire	$\dfrac{dx}{dt} = -ay$ $\dfrac{dy}{dt} = -bxy$	Brackney (1959) $\dfrac{b}{2}(x_o^2 - x^2) = a(y_o - y)$ Mixed Law
Operational losses vs. operational losses	$\dfrac{dx}{dt} = -ax$ $\dfrac{dy}{dt} = -by$	Peterson (1953) $b \ln \dfrac{x_o}{x} = a \ln \dfrac{y_o}{y}$ Logarithmic Law
Aimed fire plus operational losses	$\dfrac{dx}{dt} = -ay - bx$ $\dfrac{dy}{dt} = -bx - ay$	Morse and Kimball (1951) (Generally very complicated)

Figure 6.3 Various functional forms for attrition.

Comparison of the differential equations in Figure 6.3 with the information control differential equations (6.1) through (6.4) formulated earlier reveals striking similarities in form. Accordingly, many of the solutions and mathematical techniques developed as part of Lanchester combat theory are directly applicable to the solution of problems in information warfare control theory.

The Helmbold-type combat equations presented below represent a general combat model that contains many of the classic homogeneous force combat models as a special case.

$$\dot{x} = -a(t) \left(\frac{x}{y}\right)^{1-w_y} \cdot y \tag{6.23}$$

$$\dot{y} = -b(t) \cdot \left(\frac{y}{x}\right)^{1-w_x} \cdot x \tag{6.24}$$

6.6 COUPLING BETWEEN INFORMATION MEASURES AND COMBAT THEORY EQUATIONS

The degree to which a combatant's performance behaves more like an aimed fire engagement or more like an area fire engagement is a function of the knowledge possessed concerning the combat situation. The more complete the information a combatant possesses with respect to the combat situation, the more closely that combatant's performance will follow the laws of aimed fire. In this case, the combatant's performance approaches a value proportional to the square of the number of units in the engagement, whereas, in the case of area fire, the performance is proportional to the number of units. In short, the relative effectiveness of the units employed by two opposing forces in an engagement situation is directly related to their relative state of knowledge, or information, governing the combat situation. This relationship can be expressed by means of the Helmbold equation, where it is postulated that perfect knowledge would correspond to aimed fire, $W_x = 1$ and $W_y = 1$. Ignorance would correspond to area fire, $W_x = W_y = \frac{1}{2}$. Misinformation would correspond to worse performance than area fire, or to the cases $W_x < \frac{1}{2}$ and $W_y < \frac{1}{2}$. Total misinformation could be represented by W_y and $W_x = 0$. Under total misinformation, a combatant fires on himself and (6.23) and (6.24) reduce to:

$$\dot{x} = -ax \tag{6.25}$$

$$\dot{y} = -by. \tag{6.26}$$

In this case the more units a combatant possesses the larger are the combatant's losses, the combatant's attrition rate being proportional to the number of units he possesses. There can be mixed information cases where the enemy's state of knowledge differs from the friendly's.

The Helmbold-type combat can be slightly modified by adding an additional term for "operational" losses (i.e., losses due to sickness, accidents, or fratricide due to misinformation). If we add terms for such losses, then equations 6.23 and 6.24 become

$$\dot{x} = -a(t) \left(\frac{x}{y}\right)^{1-W_y} y - b(t)x \tag{6.27}$$

$$\dot{y} = -b(t) \left(\frac{y}{x}\right)^{1-W_x} x - a(t)y \tag{6.28}$$

These relationships can be depicted graphically [2] (see Figures 6.4 and 6.5). Here the parameters d and e (where $e = W_x$, $d = W_y$) represent the coupling of the combatant's state of information to the combat equations; (d = friendly's knowledge, e = enemy's knowledge), $d = 1$ corresponds to total knowledge, $d = \frac{1}{2}$ to complete ignorance, and $d = 0$ to total misinformation.

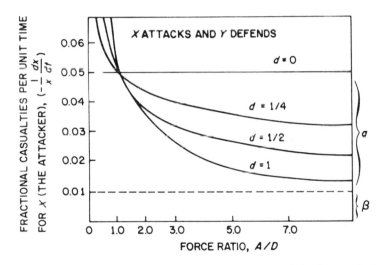

Figure 6.4 Relation between x's fractional casualty rate and the force ratio for the

$$\text{model } \frac{dx}{dt} = -a \cdot \left(\frac{x}{y}\right)^{1-d} \cdot y - bx \text{ when } X \text{ attacks.}$$

6.7 PRINCIPLES OF INFORMATION WARFARE CONTROL

By examining the differential equations that govern information warfare and combat theory and their postulated coupling, several principles emerge. They are summarized below.

• There is a threshold effect with respect to combat information. Below a given

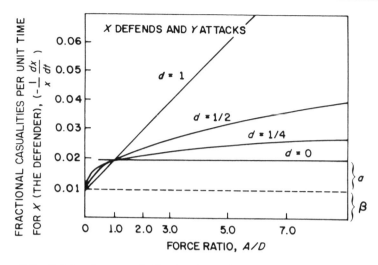

Figure 6.5 Relation between x's fractional casualty rate and the force ratio for the

$$\text{model } \frac{dx}{dt} = -a \cdot \left(\frac{x}{y}\right)^{1-d} \cdot y - bx \text{ when } X \text{ defends.}$$

threshold, added information has little effect on combat outcome and above a critical threshold added information is superfluous.

- There are time windows within which an information advantage is critical and outside of which the same information advantage becomes irrelevant. Since it is not possible to maintain an information advantage indefinitely, it is necessary to possess a good understanding of the battle dynamics and to key the timing of the information warfare measures taken to the battle dynamics.

- All information items are not equal in value. The value of a particular item of information depends on the context and the situation.

These principles have intuitive appeal. They are illustrated below with some simple examples.

6.7.1   Threshold Effect

Consider a battle situation where at the onset of conflict friendly and enemy both possess an equal number (25 units) of equivalent-type units. The value of information in this example is assumed to be reflected in the number of friendly units that can engage in aimed fire rather than area fire. The enemy is assumed to use all of its units in the area fire mode. The effect of increasing friendly's information advantage, under these assumptions, is illustrated below.

It can be seen from Table 6.1 that once the information advantage reaches a level where at least four friendly units are engaged in aimed fire, further increases in the information advantage have no effect on battle outcome; it is time to strike. Similarly, there is a lower-level threshold effect in that increasing the

Table 6.1 Threshold effect.

Initial Friendly Assets	Initial Enemy Assets	Remaining Assets
Aimed/Area	25	Friendly/Enemy
0/25	25	0/0
1/24	25	0/0
2/23	25	2/0
3/22	25	8/0
4/21	25	12/0
5/20	25	20/0
6/19	25	20/0
.	25	20/0
.	25	20/0
.	25	20/0
25/25	25	20/0

information advantage from no units to one unit engaged in aimed fire has no effect on battle outcome. The information advantage has to increase to the point where two or more units are engaged in aimed fire before there is an effect on battle outcome.

6.7.2 Time Windows

To illustrate that there are time windows or points in a battle when an information advantage is more critical than at other times or points in the battle sequence, let us consider the following simple example. Both sides are assumed to have 25 units each. The attack is initiated either when friendly possesses a high information advantage (23 units in aimed fire mode and 2 units in the area fire mode) or when friendly has a much lower information advantage (2 units in an aimed fire mode and 23 units in an area fire mode). The timing of the attack relative to the information advantage is critical to outcome of the battle. If the battle is initiated when in the high information advantage mode, friendly will completely destroy the enemy with 12 of friendly's own units remaining (almost half of his force). For the low information advantage case, friendly will completely destroy the enemy, but only two of friendly's units will remain—a bitter price to pay for victory!

The second example makes use of the relations shown in Figures 6.4 and 6.5 that relate fractional casualty rate to force ratio for the Helmbold-type combat modified to take "operational" losses into account. The curves are computed with typical values assumed for the attrition rate coefficients a and b. The parameter d (recall the previous discussion) is related to the combatant's information state, where $d = 1$ represents complete information. Consider the following two-stage battle, each stage taking one time unit. Initially, the force ratio is 125 attacking units to 25 defending units, or $A/D = 5$. Complete information, $d = 1$ or aimed fire, is assumed for the first stage of battle and complete ignorance, area fire, or $d = 1/2$, is assumed for the second stage of battle. For this case, using Figures 6.4 and 6.5, the two-stage battle results in an attrition ratio of loss-to-attacker/loss-to-defender of 3 to 1. If, on the other hand, the first stage is fought with area fire and the second stage with aimed fire, then the attrition ratio increases to 6 to 1.

It is seen from the above that there is a distinct preference for starting with an information advantage at the beginning of the battle rather than achieving this advantage during the course (in this case at the middle) of the battle.

6.7.3   Battle Dynamics

The differential equations (6.1) through (6.4) model information warfare as a time-varying process. For example, (6.4) reflects the fact that if no information-manipulating functions are actively applied, the information state will not remain static but will degrade exponentially with time. This situation is compounded by the introduction of enemy information-warfare actions. Taking all of these considerations into account, it is clearly not prudent to assume that an information advantage can be maintained indefinitely.

6.8   FUTURE RESEARCH AREAS

It should be noted that many research issues and areas of investigation remain to be pursued. Some of these issues are briefly summarized below.

6.8.1   Coupling to Lanchester's Combat Theory Equations

Much more research needs to be done with respect to examining the best means of modeling the coupling of the state of information (entropy measure) to the combat equations. For example, is the mapping between the information entropy and the combat equation parameters (d, e, or w_x, w_y) a linear or a nonlinear relationship? Is there some measure of the state of information that is better to use than the entropy? Can the assumed coupling between information state and combat equations be experimentally verified? Are the assumptions regarding the contribution of information to the coupling parameters valid? How do we account for the context-sensitive aspect of the value of a particular item of information? These and other such issues need to be addressed in any future work.

6.8.2   Sensitivity of Strategy to Objective Functions

The choice of objective function is a particularly important consideration. Alternative objective functions should be identified and evaluated. The sensitivity of the various factors (i.e., the choice of objective function, of the candidate control strategies to be considered, and of the optimization alternatives themselves) to the effectiveness and performance of the proposed information warfare system needs to be more thoroughly studied and addressed.

6.8.3 Experimental Validation

The information warfare concepts presented the exact form of the differential equations that govern the information warfare process; the coupling of information measures to the equations of combat needs to be validated by experiment.

6.8.4 Computational Complexity

There is a real issue with respect to the viability and affordability of implementing the algorithms, hardware devices, and software programs required to compute estimates of entropy information states, to generate and evaluate candidate strategies for information warfare control strategies, and to display and present these results to the decision maker.

6.8.5 Applications of Artificial Intelligence to Military Communications

As was previously discussed, communications systems, in general, and military communications systems, in particular, are faced with a bewildering array of conflicting stresses and challenges that must be met in order to operate under future battlefield conditions. The systems must operate under degrading conditions characterized by high-power sophisticated jammers, EMP, smoke clouds, and chaff. At the same time, they are being called upon to support the exchange of ever-increasing volumes of information among an increasingly distributed and highly mobile force. Furthermore, the communications information exchange must be secure and nonexploitable.

Many advances have been made with respect to available communications bandwidth, antenna gain, diversity of media, and distributed networking and routing. Although all these advances help in meeting the goals stated above, there are certain unavoidable tradeoffs and limitations. Increased resistance to jamming and errors can be achieved only through increased redundancy in transmission (e.g., pseudorandom coding), bandwidth diversity (e.g., spread spectrum, frequency hopping), and antenna gain (e.g., antenna pattern nulling). Consequently, robustness and jam resistance can be obtained only by trading against the available channel capacity and connectivity. Likewise, reduced probability of intercept and exploitation is achieved through similar tradeoffs against available channel capacity, connectivity, and reliability. Increased connectivity over wider, more dispersed areas at maximum achievable information rates and high transmitted power levels improves communications utility and reliability but also increases the chances of exploitation. Communications security measures (COMSEC) serve to further drive up the cost of communications and add to the complexity of system operation and maintenance.

With current technology, communication is both slower and more costly than computation. That is, bits can be created faster and cheaper than they can be shipped over substantial distances. Present trends indicate that these imbalances in speed and costs of communication and computation will not only continue but are likely to greatly increase. Because of these trends, there is a motivation to look for digital processing solutions that can overcome the communication design challenges and limitations previously identified. These solutions should trade off computation and communication. Ideally, to effect an information exchange the communication nodes should spend the bulk of their time computing and only a small fraction of their time communicating with one another. Let us next explore the compute-intensive field of artificial intelligence as a potential source of many of these communication-computation design tradeoffs.

To help determine the best ways to apply this technology it is useful to study how humans communicate, particularly when they are under stress and/or unusual circumstances. In this regard it is noted that under the right set of circumstances an enormous amount of information can be exchanged between two or more individuals with a minimum of communication actually taking place. A nod, a gesture, a knowing glance, a few seconds worth of disconnected, seemingly vague or innocuous words oftentimes result in an exchange that normally would have required half an hour of conversation or more. Further, these short, abbreviated communications can be conducted in the presence of third parties with knowledge of the exchange kept incomprehensible to them. The bidding process in contract bridge is one example of such an exchange. How are these conversations managed? What special techniques are the speakers employing? Are there any unique characteristics of these techniques or properties that can be applied to the military communication problem?

For one thing, these abbreviated conversations appear to be very dependent on the situation and context. The conversationalists usually belong to a special affinity group that share some experiences, specialized knowledge, or prearranged signals and codes in common that are not known by the third parties. What seems to be taking place in these abbreviated conversations is that the communicating parties are relying upon their specialized private knowledge and experience to expand, resolve, and interpret the sparse and ambiguous communications received into more detailed and meaningful concepts, themes, and intentions. The parties expend more time and effort in this interpretation phase in exchange for a reduction in the actual data that is physically transmitted. This technique is directly analogous to the digital-processing design objective outlined earlier and is a natural application for such artificial intelligence technologies as natural language processing and expert systems.

A brief example is provided below so that the reader can more fully appreciate how these technologies can contribute to the reduction of the required communications overhead. Consider the application of natural language processing to data base queries. Certainly, the requirements to access remote data bases is a common military communications requirement. The original motivation behind natural language data-base query research was not for communications bandwidth reduction; rather it was to develop an easier, more natural human interface with computer data bases. By ''natural language'' we mean languages more like the ordinary languages that evolved as the normal means of communication among people, such as English, French, Chinese, rather than the formal or artificial languages that people have invented for special purposes, such as FORTRAN, LISP, ADA, SEQUEL, and arithmetic and musical notation.

Natural language tends to be ambiguous in the sense that the same sentence can often be interpreted in many different ways. These ambiguities tend to get resolved by people through an interpretation process that makes use of pertinent background contextual information. Artificial, more formal, languages prevent these ambiguities by means of more restrictive use of the meanings, placement, and acceptable words and symbols in a sentence or phrase. Special symbols and brackets are also used to clarify the meaning and intent of a phrase. These restrictions tend to result in a longer string of characters being required to represent the same query in a formal artificial language than is required for a more natural language. This imbalance is further accentuated if the question or query asked of the data base required more than a simple, direct access of a stored item of information. Some queries are more complex, requiring several accesses of various items of data as well as some ancillary combining and processing of, and drawing of inferences from, the retrieved information in order to provide the desired information back to the requester. Here some knowledge or expertise is required of the respondent.

When using a formal artificial language, all the necessary steps, including data retrieval and processing, need to be explicitly spelled out in the data request. When a more natural language is employed, only the desired information need be requested. Not all of the necessary intervening steps need to be spelled out. The natural language interpreter, possessing knowledge of the subject matter, converts this top-level request into the necessary steps, data searches, manipulations, and inferences required to obtain the desired information and translates these steps into the appropriate artificial language that the computer understands and responds to. Figure 6.6 illustrates the kind of savings possible in information exchange overhead through the employment of such a natural language database query in place of a more traditional artificial language. In this example [3], the compound English-like query "Where is the fastest U.S. carrier within 500 miles of Naples?" requires a total of 59 characters, which translates to over 792 characters worth of commands, procedural statements, and queries in a high level, but artificial, data base language called DATA COMPUTER. This saving equates to better than a 10 to 1 reduction in the characters required to be exchanged to express the information request.

This natural language query could be still further abbreviated if knowledge of previous requests, collateral information, or displays were shared by the communicating systems. For example, if a previous question asked what the closest city was to an observed enemy platform, such as a detected Soviet nuclear submarine, and the answer was Naples, the question posed in Figure 6.6 could be reduced to "Where is the fastest carrier within 500 miles?" If a common geographic display were assumed, the question "where is the fastest carrier?" could be asked with respect to a displayed region of interest, namely a 500-mile arc around Naples. Furthermore, if a carrier symbol were pointed to on the screen, even the word "carrier" could be deleted for a more abbreviated symbol.

```
Question:
WHERE IS THE FASTEST US CARRIER WITHIIN 500 MILES OF NAPLES?

Query:
FOR NSTDPORT1, XX1 IN PORT WITH (XX1, DEP EQ 'NAPLES') STRING1 = XX1.PTP :

BEGIN DECLARE Y1 STRING (,100) ,D=')'
      DECLARE Y2 STRING (,100) ,D=')' Y2 = '00.0'
      DECLARE Y3 INTEGER Y3 = 0
      FOR XX1 IN SHIP WITH (XX1.TYPE2 EQ 'V')
                      AND (XX1.TYPE1 EQ 'C')
                      AND (XX1.NAT EQ 'US')

          FOR XX2 IN TRACKHIST WITH(
GCDIST(XX2.PTPX , XX2.PTPNS , XX2.PTPY , XX2.PTPEW , 4446 , 'N' , 1430 , 'E')
          LE 500)
               AND (XX2.UICVCN EQ XX1.UICVCN)

          BEGIN Y1 = XX1.MCSF
                IF Y1 LE '99.9'
                   AND Y2 LT Y1
                THEN BEGIN Y2 = Y1
                           Y3 = 1

          END
```

Figure 6.6 Sample DBMS dialogue (data computer).

```
END XF Y3 EQ 1
   THEN FOR XX3 IN SHIP WITH (XX3.TYPE2 EQ 'V')
                        AND (XX3.TYPE1 EQ 'C')
                        AND (XX3.NAT EQ 'US')
                        AND (XX3.MCSF EQ Y2)
           FOR NSTDPORT1, XX4 IN TRACKHIST
              WITH (
GCDIST(XX4.PTPX, XX4.PTPNS, XX4.PTPY, XX4.PTPEW, 4446, 'N', 1430, 'E')
                    LE 500)
           AND (XX4.UICVCN EQ XX3.UICVCN)
           BEGIN STRING 1 = XX4.PTD
                 STRING 2 = XX4.PTP
                 STRING 3 = XX3.NAM

                 END
```

Figure 6.6 (Continued).

The same sort of savings experienced for the data query could also be realized for the data reply by transmitting the reply back in a natural language form that the user system interprets in terms of the context and the background of the situation. These English-like queries could be still further abbreviated through various standard data compression codes and table look-up techniques. Moreover, such natural language sentences could also be made unintelligible to an eavesdropper through their translation to another private (perhaps still further abbreviated) language (new vocabulary and grammar) prior to its transmission. Although this approach would not substitute for the need for data encryption, it would provide still another level of data security. Incidentally, if the private language were sufficiently constrained to a narrow enough context, speech input and speech response could be integrated with the text and graphics interaction.

To realize such savings, however, requires some rethinking of how to interface the natural language interpreter to a system of distributed data bases. Traditionally, natural language front ends to multiple remote data bases have been envisioned as being implemented in the manner illustrated in Figure 6.7. In this manner, a user can request information in a natural language mode that is transparent to the peculiarities and artificial restrictions imposed by each individual data base system. The natural language request is then interpreted once, at a single point, where the relevant data base items and processing steps are identified, generated, and sent to the appropriate data base system(s).

Natural language processing is a very computer-intensive process and this single-point approach conserves computing power. Existing natural language processing systems typically require a large dedicated computer to interpret and translate in real time. If a natural language, distributed data-base system were implemented as illustrated in Figure 6.8, the required communication capacity

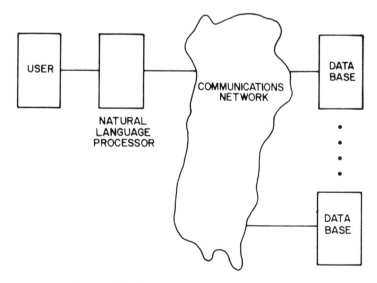

Figure 6.7 Computation resources conserved.

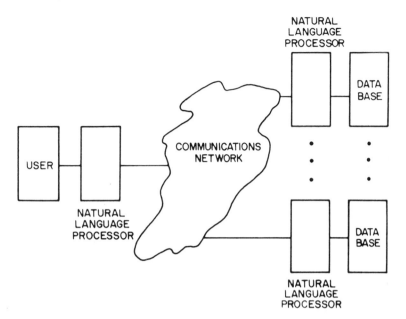

Figure 6.8 Communication resources conserved.

could be reduced to the smaller exchange of characters associated with the natural language query as opposed to the longer artificial language query of Figure 6.7. The approach of Figure 6.8 does require multiple natural language processors front-ending each data base system so that the received natural language queries can be interpreted. It also requires a natural language processor back-ending the user terminal to help determine which data base systems to send the data request to. However, if the current trends continue and computers decrease in cost and increase in performance at the current rates, this added computational burden and expense (natural language processors at each node) will certainly be well worth the savings gained in communications bandwidth and in increased security. Further, this added computational ability at each node helps to resolve ambiguities and errors and conflicts in the received data by means of previous background knowledge and inferencing mechanisms used to check for logical consistency in the received data. This ability has the potential of increasing the system's overall robustness and tolerance of communication errors, both random or intentional.

The approach illustrated in Figure 6.8 does suggest that a new class of higher-level protocols would have to be agreed upon before two or more parties could communicate, namely the agreed-upon vocabulary, grammar, and common contextual background knowledge. These agreements could be prestored and called up during initialization by means of short code words.

Distributed command-and-control data bases need to be updated and kept current. Artificial intelligence techniques can also be applied to reduce the associated communications overhead for this update process. For example, consider the need to keep cooperating units mutually informed with respect to the prevailing tactical situation. This requirement involves the exchange of information concerning the location, motion, and identification of friendly, enemy, and neutral combat units in a battlefield area of interest. Current approaches to this problem require sending updates of observed object positions and identifications in real-time as sensor detections occur. Because the situation is dynamic and subject to change, the tactical area must be kept under continuous surveillance. If the area and number of objects of interest are large, the volume of sensor observations per unit time that must be communicated, exchanged, and analyzed among participating units can represent a formidable communications load.

One way to reduce this communications load is through the creation and use of large knowledge bases (e.g., frame-based knowledge representations). Such data bases are oriented around the various objects of interest. Each object-oriented data frame will typically contain, in addition to actual sensor measurements such as features, parameters, and location, information that reflects the object's rules of behavior and default parameters that are typically (or historically) associated with that particular object. Furthermore, many observed object actions trigger associated events and actions in other objects. These interobject linkages are reflected by embedded procedures or "demons," stored in each object information frame, that get activated by the appropriate values and data updates. Past observation can be combined with these rules, default values, and embedded procedures to enable one to infer likely future activities and motions in the absence of any further observations. When new observations merely reflect actions that could have been inferred by the existing-knowledge data base, there is no need to update or modify the existing data bases. Only when new information is received that would tend to modify or contradict the existing-knowledge base is there any need to update data bases. These updates need only reflect significant changes, not information that is already known or that can be inferred from the existing-knowledge base.

In this case, memory and processing overhead is traded off for communications bandwidth. We add memory-dependent data, such as past observed patterns and object interdependencies, and the processing capacity to draw inferences and make predictions based upon this data, to reduce the required frequency of data base update. When the receiver data updates are very noisy and unreliable, the need for a memory-dependent data base is justified, not just to save communications overhead, but also to increase the quality and the reliability of the inferred tactical situation. Received sensor detections can be validated and checked for consistency and logical plausibility. Gaps and holidays in sensor measurements and observations, as well as a reduced rate of update, can be more easily tolerated while still maintaining a reasonably coherent picture of the tactical situation. Noisy, unpredictable data will usually require more frequent updates to maintain a current data base than more reliable, predictable data; however, use of knowledge base concepts should help to both minimize the need for increased data updates and keep the data base less vulnerable to corrupting effects of noise and errors in data.

Artificial intelligence techniques can be applied to many other areas of communications. They can be applied not only to reduce the required communications capacity for a given military information exchange application but also to improve system robustness and tolerance of errors and failures.

Expert system technology can be used to assist in the control and maintenance of a communications system. System control and maintenance is generally a very manpower-intensive process requiring skilled experts with extensive training and a considerable experience base. Generally, these experts, consciously or unconsciously, apply rules of thumb to diagnose and to detect abnormal, faulty communications operations requiring repair or some other remedial action. Recommended remedial actions are also derived from these heuristic rules of thumb. Expert system technologies such as production rule systems allow one to acquire and represent these rules in computer-compatible form, as in "If, Then" statements. These systems also include master control programs that determine the order in which rules should be applied to the monitored system performance to arrive at appropriate system control and maintenance actions.

The technology we have described is already being successfully applied to similar applications, such as the design and maintenance of computer systems. Such expert systems are used in two modes, both as an intelligent assistant to the expert, amplifying the capacity and quality of the expert's work, and as a surrogate for an expert when the expert is not available. The systems typically consist of some 2,000 plus rules. This size system is implementable with today's computer hardware and should prove sufficient for a large class of communications control and maintenance applications. Other areas where this technology could be applied are in training and software development.

6.9 REFERENCES

1. Moose, P. *A Dynamic Model for C³ Information Incorporating the Effects of Counter-C³.* December 1980, NPS-62-81-025.
2. Taylor, J. G. *Force-on-Force Attrition Modelling.* Military Applications Section, Operations Research Society of America, January 1980.
3. Grosz, B. "TEAM: A Transportable Natural-Language Interface System." Conference on Applied Natural Language Processing, Santa Monica, Calif., pp. 39–45, February 1983.

6.9.1 Further References

1. Barr, A., Cohen, P., and Feigenbaum, E. *The Handbook of Artificial Intelligence.* Los Altos, Calif.: William Kaufmann, 1982, 3 volumes.
2. Granger, R., Staros, C., Taylor, G., and Yoshii, R. "Scruffy Text Understanding: Design and Implementation of the NOMAD System." Conference on Applied Natural Language Processing, Santa Monica, Calif., pp. 104–106, February 1983.
3. Lehnert, W., and Schwartz, S. "Explorer: A Natural Language Processing System for Oil Exploration." Conference on Applied Natural Language Processing, Santa Monica, Calif., pp. 69–72, February 1983.

Chapter 7

C³I MILITARY EXPENDITURES AND THE DEFENSE INDUSTRIAL BASE

Continued deficits and high interest rates are on the minds of Americans daily. Questions are constantly being asked about how we can improve productivity, reduce deficits and interest rates, and still keep a strong military posture. Some say spend more. Some say spend less. Others say become more efficient. Who has the answer?

This chapter will highlight some of the problems that we have been witnessing in expenditures and productivity within the military communications community. An attempt is made to identify solutions for optimizing C³I expenditures in communications based on accepted techniques within the DoD and the economic community regarding investment, productivity, quality, and state-of-the-art technology. The implications are that the United States must reassess how it is going to spend money for the maximum benefit of the *country*. Some may ask, why such a chapter in a book on communications? The answer is that only through the wise expenditure of funds in *communications* will an effective force multiplying effect be achieved in future systems [1].

7.1 THE INDUSTRIAL DECLINE

Every indication points to the fact that the United States is now, and has been for some time, tending toward losing its dominant position as the leading industrial power. The problem seems to be that the United States has not utilized its industrial capacity to the fullest extent possible.

Recent studies indicate a declining productivity and quality. This trend has adversely affected the U.S. military posture [1]. Reduced investment in new and modern equipment as a percent of GNP and a slowdown in the rate of productivity growth have gone hand in hand with a reduced rate of spending, in constant dollars, for new and better tools for production.

Contributing to the deterioration of the U.S. industrial base is the fact that investments in research and development, as a percentage of GNP, declined dramatically over the past several years. The ratio of national (military and civilian) R&D expenditures to GNP decreased nearly 24 percent from 1964 to 1982. In comparison, the growth in R&D spending as a percent of GNP for the Soviet Union during the same period rose by 21 percent. This situation has been improved recently but may occur again.

In addition to the problem of lagging R&D investments are slips in product quality, resulting in serious consequences for the national defense posture. For example, a study [2] of certain electronic components, memory chips of a given type, that were supplied by three Japanese and three U.S. firms showed that the failure rate of the best America model was nine times higher than that of the best Japanese model.

The United States is in danger of losing its industrial preeminence. There is ample evidence to support this statement, as will be discussed. U.S. military communications systems are also in danger of losing preeminence.

Part of that evidence is the fact that, in the past two decades, productivity gains have lagged significantly behind those of other industrialized countries. According to the Bureau of Labor Statistics, productivity growth rates for the total U.S. economy and for the manufacturing sector are the lowest of all major free-world industrial nations.

The fact that, for the past 20 years, Japan's productivity growth rate for the total economy averaged over 7 percent, Italy's 5 percent, France's and West Germany's 4 percent, the United Kingdom's 2.3 percent, and Canada's 1.9 percent, while that of the United States has averaged only 1.5 percent, is part of that evidence.

The United States has truly been in an industrial decline relative to the rest of the industrialized nations. That relative industrial decline has brought us to a point today where much of our industry is less efficient than Japanese, German, French, or Italian industry, to the point where most of our consumer electronics like TVs, radios, and stereo equipment, electronic watches, and games are imported from foreign sources; to the point that recently Japan produced more automobiles and more trucks than we did. In what is probably the most automated plant in the world, Datsun produces 1,300 cars a day using only 67 workers. How? With robots. Japan has four times as many sophisticated reprogrammable robots as the United States.

The United States has been brought to the point that its number one export is the U.S. dollar.

Japan is not the only competitor. In fact, in many industrial fields, several other countries, such as West Germany, Italy, and Sweden, are ahead of the United States, and the gap is widening.

Few industries have been exempt from this erosion of U.S. industrial power, and even those sectors with tremendous trade surpluses have been losing their share of the world markets. These industries include research-intensive and technological trailblazers such as electronics, chemicals, and machine tools. Other U.S. high-technology companies such as pharmaceuticals and aircraft have also become less competitive. In 1962, U.S. pharmaceuticals had 27.6 percent of the world market share, today only 15 percent.

The United States is still competitive in the aircraft industry, but its world market share even there decreased from 66 percent to 58 percent during the decade of the 1970s. These industries are the main contributors to our technological edge in defense systems and are the very core of our defense industrial base. The continuing loss of market share has major security implications because the loss of markets also means the loss of capacity, capability, and skills that could be essential during a national cmcrgcncy.

The United States was once the envy of the world for its ability to find new ways to make things better, with better quality and at lower costs. It is now lagging in increasing the amount of goods and services produced per employee when compared with a number of other nations. This failure to improve productivity growth is one major reason that prices are increasing and the United States is becoming less competitive and a debtor nation.

In the 1950s and early 1960s, the United States enjoyed a relatively high productivity growth rate. During this period, the Consumer Price Index (CPI) increased only 2 percent per year for all goods and services. Companies were investing in new and better plants and equipment and, more important, the standard of living was constantly rising. However, during the late 1960s and early 1970s, the CPI increased significantly. The more recent performance from 1979 to 1982 shows productivity gain slipping badly, at the same time that the CPI has increased dramatically, averaging 10.2 percent per year, as indicated in Figure 7.1.

As for the impact on the field of communications, the lag in R&D expenditures and the decline in quality production have adversely affected U.S. command, control, and communications (C³) capability. Many research institutes have indicated concern that U.S. industries do not have the capability of responding, with quality productivity, even if R&D and investment dollars were spent in improving C³ capability.

In order to overcome the majority of deficiencies in C³ from 1986 through 2000, approximately 40 billion dollars would have to be spent. This figure represents an unconstrained (more rapid rate of spending) budget in which the majority of R&D budget is spent in the 1986 through 1990 timeframe. Since U.S. spending in R&D has been at a decreased rate for a long period of time, it appears that this accelerated expenditure is required to improve C³ capability. The question is: Even with an accelerated expenditure in R&D and investment, will the United States really achieve an improved C³ capability? Previous history and recent studies indicate that the answer to this question is no! The R&D expenditures in C³ over the last decade have not resulted in significant improvements in capability. Only a very low percentage of the R&D money spent has resulted in improved communications capability. For example, despite serious deficiencies in communications connectivity, very little has been done to improve the situation. Very little of the R&D funds spent have actually reached industry for production. Consequently, the industrial base has not had to respond to productivity by maintaining high-skill personnel and production lines. Much of today's R&D is spent in complying with government regulations. Compliance

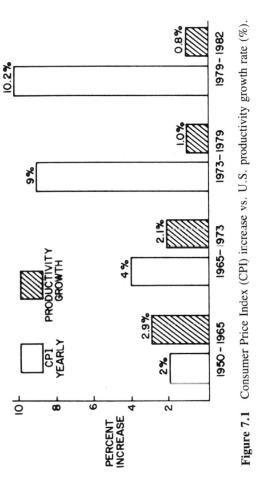

Figure 7.1   Consumer Price Index (CPI) increase vs. U.S. productivity growth rate (%).

procedures siphon off as much as 40 percent of total R&D investments in some industries. Furthermore, continually changing requirements and inadequately defined specifications reduce the efficiency with which R&D money is spent.

The challenge for the future is to expend R&D and investment dollars in such a manner as to increase C^3 capability and industrial capacity. All indications are that the two go hand in hand. Accelerated expenditures focused in the correct manner will improve the U.S. C^3 posture as well as its industrial capacity and production rates.

7.2 PRODUCTIVITY AND QUALITY

The productivity of U.S. industry has declined significantly relative to that of foreign competitors. The United States has become less competitive in foreign markets and in certain domestic markets. This decline, in turn, has reduced the capital available to invest in productivity improvements. One reason for this condition is the decrease in expenditures for R&D in recent years. The amount of new industrial technology that can be used by U.S. companies to become more productive and competitive has been reduced as a result of these lower funding levels. Shortages of skilled labor, scientists, and engineers further exacerbate the problem.

Finally, the low quality of many U.S. products has further contributed to the decline of industrial productivity. Foreign competitors, notably the Japanese, are taking over a growing number of U.S. markets. Their success is largely attributable to knowledge gained from the United States and from a long-term commitment to quality products. Furthermore, these successes have been duplicated in Japanese plants located in the United States and using American workers, a fact that strongly suggests that U.S. leadership and management commitment to quality could be far more extensive than it is at present. Profound change is needed if the United States is ever to expect to recover the ground it has lost.

As lesser-developed industrialized countries combine high technology with cheaper labor to produce and sell everything from steel to high technology electronics and aircraft, they are closing the gap between themselves and the United States in producing more per worker.

The U.S. Council of Economic Advisors projected a 1.5 percent economic growth rate for the United States over the next ten years or so. From recent evidence, that forecast looks very, very optimistic. But even if the United States did have the 1.5 percent growth rate, the Council projected that France would overtake the United States in total worker productivity in 1986, Germany in 1987, and Japan in 1988, with some others not far behind.

Although the fact of stagnating U.S. productivity is crystal clear, the factors responsible for this stagnation are not that clear. Most researchers, however, have focused on capital investment in modern plant equipment and research and development as the principal prerequisites for productivity growth.

7.3   INVESTMENT IN NEW PLANT AND EQUIPMENT

Today the average U.S. plant is 20 years old—eight years older than the equivalent German plant and more than 10 years older than the equivalent Japanese plant. This difference results from the fact that Germany and Japan are reinvesting more of their GNP than the United States is. The percentage of GNP invested in capital goods by industrialized countries as a function of average annual productivity growth rate is shown in Figure 7.2.

The United States is last among industrial countries in investment in new and modern equipment as a percent of GNP (see Figure 7.3). Furthermore, a significant share of capital investment in recent years has been for the installation of pollution abatement and health and safety equipment. A report by the Joint Economics Committee, Congress of the United States, stated: "The low capital stock resulting from the recent inadequate levels of investment is considered by many economists as the number one cause of the productivity slow down." The committee recommended that further steps be taken to strengthen investment, including policies that will increase business and fixed-capital investment. The suggestions of the Committee still hold true today.

7.4   INVESTMENT IN TECHNOLOGY

For many years, the United States invested heavily in research and development and held undisputed first place in the world. Contributions of high technology industries to the U.S. economy have been great because of a continuous and deliberate effort to increase productivity through the technological innovation spawned by research and development. The results of these past R&D expenditures have provided the base for U.S. economic growth and the capital for future investment.

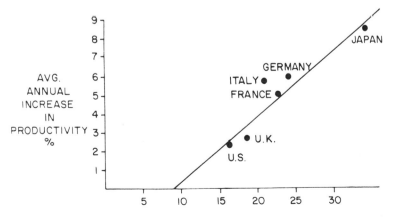

Figure 7.2   Ratio of investment to GNP vs. productivity growth (manufacturing 1960–1977).

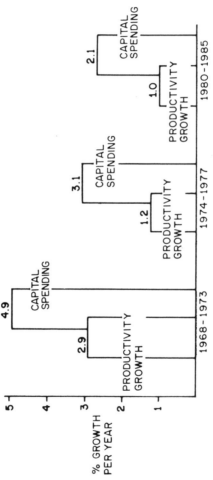

Figure 7.3 U.S. productivity growth vs. capital spending growth for tools of production.

COMPUTER SCIENCE PRESS, INC.

1803 RESEARCH BOULEVARD
ROCKVILLE, MARYLAND 20850 U.S.A.

(301) 251-9050

SAN # - 200-2361

ANNOUNCEMENT

Concerning the book U.S. Military Communications: A C³I Force Multiplier by
F. J. Ricci and D. Schutzer, Computer Science Press, Inc., Rockville, MD, 1986,
the publisher and authors wish to announce that major portions of section 8.9.19
(pages 227-237) have been taken verbatim from NRL Report 8637, "Preliminary System
Concept for an HF Intra-Task Force Communications Network" by J. E. Wieselthier,
D. J. Baker, A. Ephremides and D. N. McGregor, August 9, 1983, without prior
consent of the authors of the report. The publisher and authors express their
regret for this occurrence.

Despite the fact that technical innovation can secure large benefits to the society, and despite the fact that industrial innovation is the primary means of improving productivity, as we have noted, U.S. investment in R&D as a percentage of GNP has seen a significant decline. Increased R&D spending does not ensure an increase in the world's scientific knowledge, but it does represent an increased national scientific and technical capability. Actual expenditures by the Soviet Union have exceeded those of the United States (see Figure 7.4).

The figures tell only part of the story about industrial R&D spending. There have been disturbing indications that point to a decrease in the rate of innovation and in that portion of the R&D investment devoted to new product lines and basic research. The final report of the President's Advisory Committee on Industrial Innovation in 1980 noted that much of today's R&D is spent in order to comply with governmental regulations and to upgrade existing technology. For some industries, compliance procedures were averaging up to 40 percent of the average national R&D expenditures. This does little to aid productivity.

7.5   C³I INVESTMENT RECOMMENDATIONS

Many studies have recommended techniques for improving both U.S. military expenditure procedures and industrial capacity. In testimony before the House Armed Services Committee, Industrial Readiness Panel, Alton D. Slay, former Commander of the Air Force Systems Command, indicated that the United States is "slipping and sliding on a course toward the status of a second-rate industrial power, yet we cannot have a healthy defense industry without comparable health in U.S. industry as a whole." [3] One way of improving U.S. productivity is to spend both R&D and investment dollars at a higher rate.

As pointed out in "Special Analysis, Budget of the United States Government, Fiscal Year 1981," the industrial base does not have the capacity to respond to

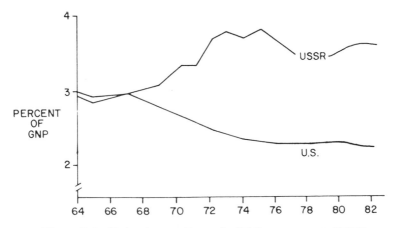

Figure 7.4   National expenditures for R&D as a percent of GNP.

the needs of the Department of Defense. That is, the present capacity of industry is sufficiently low that it would take years to respond to a surge in spending. Figure 7.5 shows a plot of relative industrial capacity, unit product cost, and DoD spending for Accelerated Budget expenditures from 1981 through 2000. Both the unit product costs and plant capacity are in normalized units.

This figure indicates that as spending increases for a fixed plant capacity and unit product cost, the inflationary factor that results from a high demand causes the unit cost of the product to increase. The effect of the peak expenditure in 1995 is not reflected immediately in a corresponding price increase. As industry responds to the demand, there is not enough capacity. Therefore, the unit price of the product goes up. The peak in the price of the product is delayed for a number of years as industry begins adjusting. New industries and subsequent capacity are then added to the system. The increased capacity and competition then begin driving the unit price back down.

7.6 PROPOSED SOLUTION FOR INCREASED RATE OF SPENDING IN C³I

In order to keep pace in the C³I arena, the United States must accelerate its investment, that is, older systems must be replaced by newer systems that meet any enemy threat and provide needed endurance. One suggested solution is to fund a C³I space-based system at an accelerated rate to make up for the recent lags. This work in part led to the U.S. Strategic Defense Initiative (SDI) [4].

It is only by innovative R&D and investment expenditures that the United States can regain its lead in C³ and solve the investment lag problem. Figure 7.6 presents a conceptual view of an innovative C³I communications system the United States could strive for during the next twenty years. This system would include (1) a survivable satellite system with sensors having communications connectivity to air, land, and sea forces and (2) a land-based system consisting of survivable communications connectivity to mobile command centers from sensor systems. Maximum advantage should be taken of new technology, in command center and communications techniques, pushing the state-of-the-art.

Optimized expenditures for a fiscally unconstrained, accelerated budget C³I architecture result in early investments in new initiatives and a significant peak in 1995. Of the total $56.2 billion fifteen-year cost, approximately 57.3 percent is spent on investment, 26 percent for O&M, and 16.6 percent for R&D, as shown in Figures 7.7 and 7.8.

The expenditure in new initiatives is $40.2 billion for fifteen years. Since this architecture has no fiscal constraints, there are no yearly or cumulative fifteen-year limits to consider. Consequently, the behavior of the expenditures profiles are somewhat more erratic than for other types of expenditures.

The yearly costs are calculated by summing the ongoing program costs plus the costs for new initiatives, minus the costs for programs deleted. The devel-

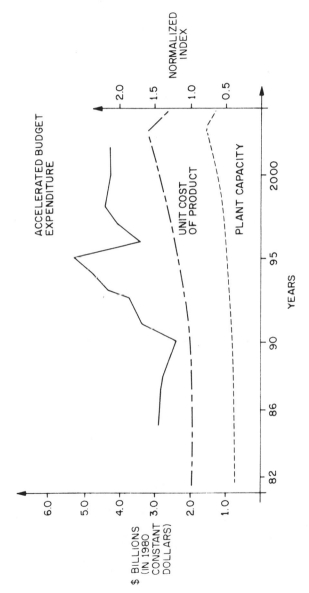

Figure 7.5 Relation of expenditures to industrial capacity and production costs.

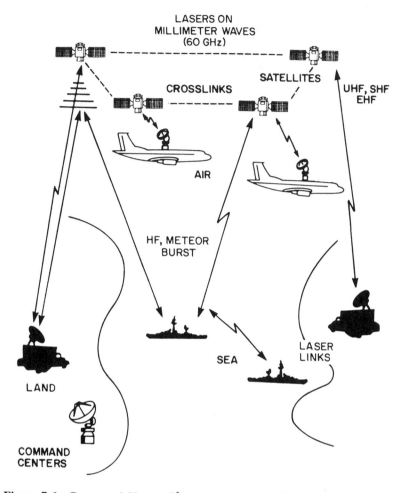

Figure 7.6   Conceptual 20-year C³I communication architecture for space-based
systems.

opment of costs involved a careful balancing of investment, R&D, and O&M
to provide a comprehensive funding profile.

The Accelerated Budget C³I Architecture introduces more new initiatives than
any of the other architectures. Also, many of the C³ systems are eliminated,
replaced, or phased down in operational capability. In particular, the following
systems would have some form of reduction in O&M expenditures after 1990:

- TW/AA
  JSS
  DEWline
  OTH-B (old)
  NAVSPASUR

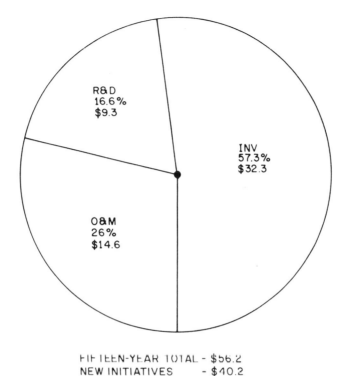

FIFTEEN-YEAR TOTAL - $56.2
NEW INITIATIVES   - $40.2

Figure 7.7   Distribution of costs of accelerated budget for architecture (in billions of constant 1980 dollars).

- Command Center
  NMCC ADP
  ADCOM ADP
  SAC ADP
  CCPDS
  WWMCCS ADP support
  AABNCP
- Communications
  TACAMO
  PACCS communications

The dominant new ''STAR WARS'' initiative added to this architecture includes:

- LWIR/Radar Satellite
- A Survivable C³ Satellite
- Survivable Space Defense C³
- Mobile Platforms
- CONUS LF Network

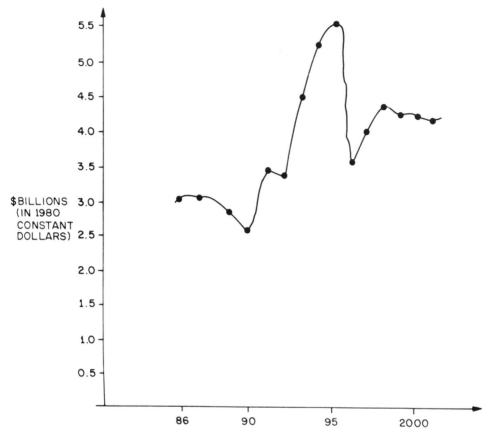

Figure 7.8 Summary of costs for accelerated budget architecture.

- Reconstitutable Satellite System
- RPV Relay
- EHF Terminals on Bombers and Sensors
- Deployment of Airborne C³
- Laser ASAT

The yearly cost for the Accelerated Budget Architectures are not constrained in any manner. Consequently, the fiscal profile does not have a predictable shape. However, in order to obtain the C³ capability required a considerable expenditure is incurred in the 1994–1995 time period. Figure 7.8 provides a plot of the costs for the ongoing and new initiative programs contained in this architecture.

Initially, about $3.0 billion per year is expended. In 1990 there is a dip, since most of the initial procurements have taken place. However, in 1995 there is peak expenditure of about $5.6 billion. This abrupt peak results from the five-year costs of the acquisition of the Long Wave Infrared (LWIR)/radar satellite, at $5.1 billion, the survivable satellite, at $1.2 billion, EHF terminals at $1.1 billion, and an RPV relay at $2.8 billion.

In 1996, there is a dip in the curve, since most of the expenditures for systems with IOCs prior to 1995 have been completed. The profile peaks again in 2000 because of the deployment of the survivable C^3 satellite at $9.0 billion, the airborne C^3 system at $1.5 billion, and the enduring RPV at $0.9 billion. All costs are for five years. Generally, the cost profile has an upward trend with a large peak in 1995.

7.7   COSTS BY FUNCTIONAL AREA

The yearly cost profile for the Accelerated Budget Architecture indicates that tactical warning and attack assessment (TW/AA) and communications expenditures are much greater than those for command centers. As Figure 7.9 indicates, the TW/AA costs have a general trend upward with a peak in 1995. The majority of the communications expenditures are from 1992 to 1995. The command center (CC) costs dip from 1991 to 1995.

As indicated earlier, the dominant costs incurred for communications systems including the RPV relay system and EHF terminals for the bombers and sensors cause the 1992–1995 peak. Even though the TW/AA costs do not peak as high as the communications costs, their total for five years is $9.28 billion, compared with $7.68 billion for communications. The dominant costs for TW/AA systems are covered by new initiatives in the LWIR/radar satellite, survivable satellite, and survivable space-defense C^3 deployment. A total of $9.0 billion is spent for the survivable C^3 satellite from 1991 to 1995.

In this architecture, most of the money for command centers is spent in the 1986–1990 timeframe for mobile platforms. A total of $1.6 billion is spent over five years. After 1990, the next major expenditure is for replacement of the mobile platforms in the 1995–2000 timeframe. Generally, this architecture results in very large expenditures in the TW/AA and communications systems to overcome the critical C^3 deficiencies.

7.8   COSTS BY APPROPRIATION CATEGORY

There is more money spent on new investments than for O&M or R&D for the Accelerated Budget Architecture. Approximately $3.0 billion or 75 percent of the new initiative costs of $40.2 billion are spent on investments. The remaining new initiative expenditure is for R&D, 20.3 percent and O&M, 4.9 percent. The low O&M costs result from the reduction in O&M costs for systems such as JSS, SPADOC, DEWline, command node ADP, TACAMO, and PACCS. For command centers alone, approximately $500 million is saved in O&M costs

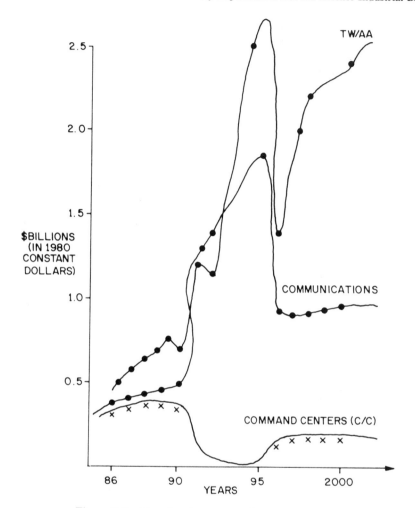

Figure 7.9 Functional area costs for accelerated budget.

from 1991 to 2000. For TW/AA systems, approximately $1.5 billion is saved in O&M from 1990 to 2000.

As Figure 7.10 indicates, initially the R&D costs are about $1.0 billion per year, but by 2000, these costs are down to about $0.2 billion since most of the new R&D initiatives required are carried out in the earlier years.

After 1991, most of the cost is for new initiatives in investment. A high rate of expenditure is increased from 1991 to 1995 in order to produce the key systems for this architecture as outlined previously.

The general trend for this architecture is a large expenditure in the 1991 through 1995 period for new initiative systems to overcome critical sensor and communications deficiencies.

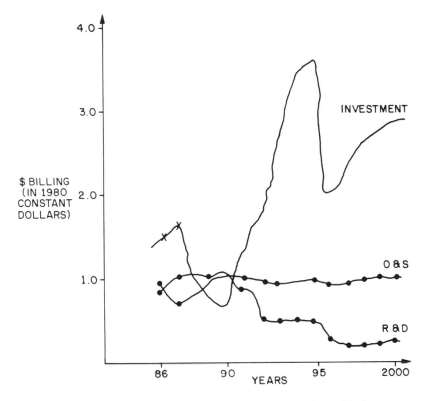

Figure 7.10   Appropriation category costs for accelerated budget.

7.9   ONE ACCEPTED INVESTMENT STRATEGY

The Heritage Foundation has published a detailed study of Defense Department requirements that placed emphasis on "quick fixes" in the C^3 strategic sector. One of the techniques it suggested that has been getting more attention in the community for weapons procurement is Preplanned Product Improvement (P^3I). This technique also has applicability in C^3I systems procurement. In general, the approach proposed is a reordering of management priorities in a sensible manner to meet the budgetary constraints that appear to be ongoing in the forseeable future.

The consensus among 250 or more senior managers of defense programs from government and industry, and the logical implication of an earlier Defense Science Board (DSB) study, recognizes that a number of attributes of success must be achieved simultaneously. These are:

- The reduction of total system costs
- Longer system lifetime before a replacement is necessary
- Reduction of initial risk

- Slower obsolescence rate of fielded equipment
- Higher operational readiness during system lifetime
- Resolution of affordability problems
- Higher technological performance during system lifetime through more rapid fielding of technology advances

Figure 7.11 shows a simple diagram depicting the accepted strategy. The diagram indicates that for a 3 percent/year annual growth, a 27-year equipment useful life can be obtained for a $30 billion 15-year expenditure. If the P³I concept is utilized to its fullest extent, early fielding of modular systems that can be easily upgraded and improved in the field would result in longer lifetimes and reduced 15-year expenditures. For example, a 40-year lifetime could be expected for a $20 billion investment.

This concept appears to have considerable merit in providing improved capability for C³I systems over the lifetime of the equipment. This procurement technique should alleviate many of the current problems of aging, obsolescent, equipment, and systems that do not adequately address a changing threat. Furthermore, this conceptual strategy for R&D investment and requirements definition should alleviate many of the problems associated with the present lag from R&D to procurement of equipment. The proper expenditures of money (up front) will be made up by improved performance and lower operations-and-maintenance costs in later years.

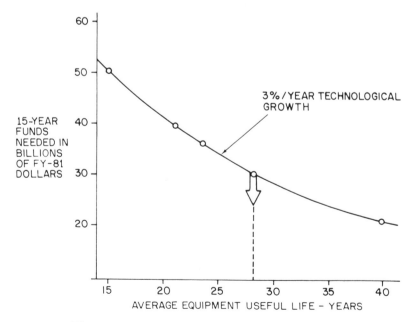

Figure 7.11 Typical costing profile for P³I approach.

Generally, a more effective military force can be achieved by *prudently* utilizing high technology investments to counter enemy threats. This fact has provided the basis for the strategy of the Strategic Defense Initiative (SDI).

7.10 REFERENCES

1. Ricci, F.J. "C³I Military Expenditures and the Defense Industrial Base." *Signal*, June 1981.
2. Committee on Armed Services. "Capability of U.S. Defense Industrial Base Hearings Before Panel on Defense Industrial Base." House of Representatives, 96th Congress, 1980.
3. Slay, A.D. "The Air Force Systems Command Statement on Defense Industrial Base Issues." Industrial Preparedness Panel, House of Representatives, 96th Congress, November 1980.
4. Delauer, R.D. The Strategic Defense Initiative (SDI). Defense Technologies Study, Department of Defense, April 1984.

Chapter 8

FUTURE TECHNOLOGY OPPORTUNITIES

The advent of small, low-cost, reliable, solid state electronic devices, including microprocessors and signal processors utilizing very large scale integrated circuits (VLSI) and very high speed integrated circuits (VHSIC) and other technological advances is bringing a new era to both military and commercial communications. The exploitation of this advanced technology will help maintain and enhance the "force multiplier" necessary to achieve a more effective C^3I system for the military.

This chapter will discuss some of these new technological advances and explore how they will impact the design, development, and fielding of military communications systems in support of C^3I. Included are such technologies as fiber optics, lasers, packet switching, spread spectrum, millimeter waves, local area networks (LANS), wideband HF, and meteor burst communications.

In the future, the military communicator will have available compact, low-cost, reliable electronic systems for voice/data communications at the VHF through EHF frequency band. Terminals will be small enough to be placed in helmets. Integrated communication, navigation, Identification-Friend or Foe, (IFF), or position determining will be commonplace. Hand-held VHF radios will be available on the battlefield to communicate over short distances and determine position location. Fiber optics and lasers are already providing new, low-cost access to wide bandwidth capabilities for land, sea, or air implementation. Millimeter wave, meteor burst, and advanced HF techniques are being developed that will provide for improved communications in critical situations.

Current systems employing some of these techniques are the Position Location Radio Systems (PLRS), Joint Tactical Information Distribution System (JTIDS), Single Channel Radios (SINCGARS), and advanced signal processors for satellites and satellite terminals. Figure 8.1 shows state-of-the-art technological advances for communication in wartime situations. Fiber optic links connect mobile platforms together in quick-reacting battlefield situations. Because of advanced technology sounding techniques under microprocessor control, HF communication, with wideband characteristics, is now practical for long dis-

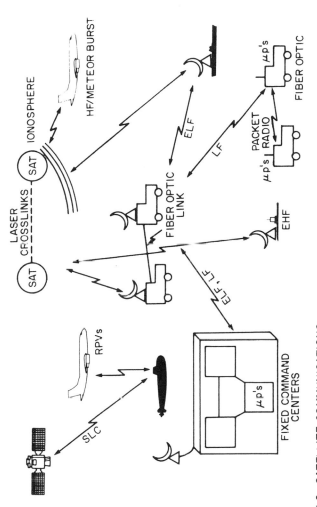

Figure 8.1 State-of-the-art technology utilized in wartime situations.

SLC – SATELLITE COMMUNICATIONS
LANs – LOCAL AREA NETWORKS
μp's – MICROPROCESSORS
RPVs – REMOTELY PILOTED VEHICLES
MP – MICROPROCESSORS

tances. Laser and millimeter wave crosslinks provide wide bandwidth links between Eastern and Western Hemisphere satellites.

EHF capability with onboard satellite processing and control can now be utilized to communicate with submarines and aircraft, taking advantage of small state-of-the-art antennas and receiver/transmitter signal processing. Fixed command centers are connected by high-speed fiber optic data links to computers and commander terminals in a Local Area Network (LAN) configuration. LANs are also being used for wideband video and data communications on mobile platforms. Microprocessors have enabled the development of low-cost compact *packet switches*, high data rate *digital radios*, satellite terminals, man-pack radio control systems, and many other communications areas. Electronics technology is providing the "force multiplier" capability to enable Army, Air Force, Navy, and Marine Corps communications to meet wartime needs.

Advances in semiconductor technology and processing capability have made it possible to perform functions that were never feasible before. For example, charge-coupled devices (CCD) can be utilized to process spread-spectrum signals. Surface acoustic wave (SAW) matched filters can be utilized to process direct sequence (DS)/frequency hopped (FH) signals.

Very high speed integrated circuits (VHSIC) wil be used to process signals at rates of 25 megahertz (VHSIC I) or greater (VHSIC II) and are finding new applications in receiver/transmitter design and EW processors.

Microprocessors and associated software have facilitated the proliferation of communications systems and made them easier to develop, use, and maintain. Array processors are enabling speeds of up to 100 megaflaps. Increased memory capacity can be achieved with high capacity read only memories (ROMs) and random access memories (RAMs). Faster clocking speeds and expanded instruction sets are enabling the communications engineer to create more innovative designs with greater performance.

High-speed, ruggedized, tempest-approved microprocessors (μp's) are available from many maufacturers for land, sea, and space-based platforms. Standardized software languages such as Ada are now available for use in software development. The use of μp's has opened up avenues for "force multipliers" as never before witnessed in the military.

Artificial Intelligence (AI) and Robotics represent another area that is having impacts in telecommunications. Artificial intelligence enables decisions to be made by "smart" systems, thus reducing errors and the system's vulnerability to enemy threat. Unmanned, remotely piloted land, sea, and space vehicles are being utilized to make the warfighting capability less risky and more responsive.

Technological advances are clearly providing the "force multiplier" required to command, control, and win battles. In the next ten years, major technological strides will take place. The key to providing effective communication systems is to imbed that technology in new developments and field them as quickly as possible.

Sections 8.1 through 8.8 will outline the new technological advances. Section 8.9 will outline future system applications of these technological opportunities, including appropriate references.

8.1   SPREAD-SPECTRUM COMMUNICATIONS

Spread-spectrum modulation is one of the more significant innovations. The use of spread-spectrum techniques makes it extremely difficult for an enemy to detect or to jam (interfere with) a transmitted signal. In addition, the received spread-spectrum signal can be used by the receiver to determine its range from the transmitter to within a few meters.

Spread-spectrum systems have been developed since about the mid-1950s. The initial applications were for designing antijamming tactical communications and guidance systems [1]. A definition of spread spectrum that adequately reflects the characteristics of this technique is as follows:

Spread spectrum is a means of transmission in which the signal occupies a bandwidth in excess of the minimum necessary to send the information; the band spread is accomplished by means of a code which is independent of the data, and a synchronized reception with the code at the receiver is used for despreading and subsequent data recovery.

Under this definition, standard modulation schemes such as FM and PCM, which also spread the spectrum of an information signal, do not qualify as spread spectrum.

There are many reasons for spreading the spectrum, and if it is done properly, a multiplicity of benefits can accrue simultaneously. Some of these are:

- Antijamming
- Anti-interference
- Low probability of intercept
- Multiple user random access communications with selective addressing capability
- High resolution ranging
- Accurate universal timing

Although the current applications for spread spectrum continue to be primarily for military communications, there is a growing interest in the use of this technique for mobile radio networks (radio telephony, packet radio, and amateur radio), timing and positioning systems, some specialized applications in satellites, etc. While the use of spread spectrum naturally means that each transmission utilizes a large amount of spectrum, this may be compensated for by the interference reduction capability inherent in the use of spread-spectrum techniques, so that a considerable number of users might share the same spectral band. There are no easy answers to the question of whether spread spectrum is better or worse than conventional methods for such multiuser channels. However, the one issue that is clear is that spread spectrum affords an opportunity to give a desired signal a power advantage over many types of interference, including most intentional interference (i.e., jamming).

The major systems questions associated with the design of a spread-spectrum system are: How is performance measured? What kind of coded sequences are used and what are their properties? How much jamming/interference protection is achievable? What is the performance of any user pair in an environment where there are many spread-spectrum users (code division multiple access)? To what extent does spread spectrum reduce the effects of multipath? How is the relative timing of the transmitter-receiver codes established (acquisition) and retained (tracking)?

In ordinary radio communications, the signal bandwidth is approximately equal to the information rate; the use of a larger signal bandwidth is unnecessary and would interfere with other signals at adjacent frequencies. In spread-spectrum communications, however, the signal bandwidth is typically hundreds or thousands of times greater than the information rate. Because the spread-spectrum signal occupies a broad range of frequencies, the signal power is spread over a large frequency spectrum and the power spectral density is very low at all frequencies.

The benefits of spread-spectrum communications include transmission and reception at high data rates, low probability of intercept (LPI), resistance to jamming (AJ), and very accurate passive (receive-only) navigation. Because of the wide bandwidth at EHF these advantages are maximized when EHF band frequencies are used. EHF frequencies provide other advantages as well: highly directional antennas may be used to further decrease the probability of interception and the effects of jamming. The use of wavelengths such as mm wave at which signals are absorbed by atmospheric gases makes it extremely difficult for distant enemy ships or aircraft to intercept the strongly attenuated signals, although the signals can reach relay satellites directly overhead with only modest attenuation.

Closely associated with spread-spectrum signals is the employment of burst communications systems, which transmit data in a high-speed bit stream. The maximum transmission duration of a bit stream, even when extensive bit error redundancy techniques are used (which modestly increase the on-the-air time), is rarely longer than 30 to 45 seconds. Thus, burst communication systems are less susceptible to detection than other forms of tactical communications, simply by virtue of their minimal transmitting period.

The means by which the spectrum is spread is crucial. Several of the techniques are (1) "direct-sequence" (DS) modulation, in which a fast pseudorandomly generated sequence causes phase transitions in the carrier containing data, (2) "frequency hopping," (FH) in which the carrier is caused to shift frequency in a pseudorandom way, and (3) "time hopping," (TH) wherein bursts of signal are initiated at pseudorandom times. Hybrid combinations of these techniques are frequently used, such as FH/DS/TH.

Each technique has advantages and disadvantages and specific applications. Table 8.1 summarizes the application, features, and processing techniques of the various techniques. In current practice, hybrid techniques are getting more attention because they have much better performance for AJ, LPI, and electronic support measures (ESM) than any of the other techniques alone.

Table 8.1 Application of spread-spectrum modulation techniques [2].

Application	Technique	Features	Average Spectrum
Anti-Jam Anti-ESM (DF/Locate)	FH-Frequency Hopping	Periodic pulses at random frequency	Long-Term-Average Spread Spectrum
LPI & Covert Anti-ESM Anti-JAM	DSE-Direct Sequence Encoding (also called "Pseudo-noise")	Random reduction in signal spectral energy density because of linear phase modulation	Short-Term-Average Spread Spectrum
Some LPI and Anti-ESM mostly used with other techniques	TH-Time Hopping	Random pulses at fixed frequency	Long-Term-Average Spread Spectrum
FH, LPI Anti-ESM	FH/DSE/TH Hybrids	Signal design gives special properties	

From a military communications point of view, it is important to design the communication system to gain the maximum possible advantage of the communications link under *stressful* situations. That is, one must maximize the AJ advantage and minimize the probability of intercept. Typically, processing gain (PG) quantifies the interference suppression process; where

$$PG = \frac{\text{Jamming Power in spread-spectrum BW}}{\text{jamming power in message BW}}.$$

The ability of a spread-spectrum technique to reduce the amount of jamming power in the message bandwidth against a certain jamming strategy is quantified by the processing gain.

In today's military systems, spread-spectrum techniques are utilized throughout the frequency range in order to optimize the *information* carrying capacity under military stress situations such as jamming and interception. Table 8.2 summarizes representative DoD AJ and LPI programs for the various services. Spread-spectrum techniques are utilized in man-pack systems such as SINCGARS as well as for packages on MILSTAR and HF radios.

8.2 METEOR BURST

Meteor burst communication is based on RF signals bouncing off ionized trails in the atmosphere. The ionized trails of electrons are caused by meteorites bombarding the earth's atmosphere. The size of each trail and its duration as a reflecting or reradiatory path of RF signals depends largely on the size of the

Table 8.2 Representative DoD AJ/LPI programs.

Principal Purpose & Technique		Sponsor	Contractor
AJ/LPI Signal Design	Sidelobe Cancellers—FSC-78, 79 Narrowband Cancellers for ECCM COMSAT Terminal Applications	Navy	Harris
AJ Signal Design	Enhanced JTIDS System Hybrid—UHF Voice	AF	Hazeltine
LPI COMSEC, Power Mgmt.	Flight-Deck Communications (UHF Secure Voice)	Navy	GTE—Sylvania
AJ Signal Design	SINCGARS FH	Army	ITT
AJ/LPI Directivity, LPI	SNAP—HF/VHF Steerable Null Antenna Processor	Army	Hazeltine/ Martin Orlando
AJ Signal Design	COMBO Radio (ARC-182) FH-SINCGARS and Have Quick Wave Forms	Navy	Rockwell-Collins
AJ Signal Design	HFIP (USQ-83, HFPM Modems, various) FH	Navy PME-110	Various
AJ Signal Design	JTIDS Hybrid DTMA Voice + Data and Rel Nav	AF/Navy JSPO	ITT, Hughes, Singer
AJ Signal Design	Have Quick FH	AF/Navy	Magnavox, E-SYS. Rockwell
LPI Power & Freq. Mgmt.	MISR (Mobile Intercept Resistant Radio)	Army CECOM	ITT
AJ Signal Design	EHF/Navy Transition Package/MILSTAR	AF/Navy SP/PME-106	Lincoln Lab
AJ Signal Design	JRSVC (Seek COMM) Hybrid—DTDMA	NSA/AF	Lincoln Lab

meteor. The height or altitude of these trails range from 85 km to approximately 125 km. These heights and the curvature of the earth limit communication to ranges around 2000 km, but there is no dead space between zero and maximum range, as is frequently the case with HF.

In a typical meteor burst system, one station continually broadcasts a "probe" signal. Receipt of this probe signal by a second station indicates the existence of a usable meteor trail and the second station responds over the reciprocal path. The lifetime of a useful path ranges from four milliseconds to several seconds. A typical trail lasts for a few hundred milliseconds, with wait times between trails ranging from seconds to minutes, depending on daily and annual cycles. Each message consists of bursts of high-speed data of tens to hundreds of characters separated by periods of silence.

There are two general configurations for meteor burst systems: full-duplex, point-to-point; or master station and some number of slave or remote stations operating in either full or half duplex mode.

The "footprint" of the signal is generally elliptical in shape with major axis in the direction of transmission (typically 30 miles major axis and 15 miles minor axis). This permits multiple use of radio frequencies and provides protection against unwanted intercept and jamming.

There are daily and seasonal variations, with the best communications usually in the morning hours and the summer season. Yearly average wait times are about 20 seconds with a .5 probability, but this can be improved through the employment of higher radiated powers and higher gain antennae. Another advantage of meteor burst communications systems is that they perform reliably in an auroral environment (northern latitudes).

Nuclear effects on meteor burst communications are scenario dependent and controversial, and much of the literature is classified. It is agreed, however, that meteor burst communications will recover more rapidly from the effects of nuclear weapons than will conventional HF; the higher the frequency, the faster the rate of recovery. Progress in solid state devices and in microprocessors has enabled design of meteor burst communications systems with far more capabilities than earlier systems.

8.3 MILLIMETER WAVES

Millimeter waves are a very promising media for LOS communications in the lower atmosphere, under battlefield environmental conditions. Millimeter waves exhibit high data rates, very narrow and hard to intercept beamwidths, and low propagation losses in the 35 GHz and 94 GHz atmospheric windows.

The current decade has seen considerable progress in the use of hybrid microwave integrated circuits (MIC). The hybrid MIC has been realized through a variety of functional components [i.e., field effect transitors (FET)] and packaging developments that have made the introduction of MM wave more practical. Two- and three-dimensional beam lead impact avalanche transit time devices

(IMPATTS), for example, reduce packaging parasitics, improve reliability, increase output power efficiencies, and limit the variation of package parameters from device to device. Other new techniques include the three-dimensional beam lead structure, microstrip hybrids using planar monolithics on low-cost substrate, as well as computer-aided design techniques that can now emulate circuit designs with excellent agreement between predicted and empirical results.

8.4   OPTICAL COMMUNICATIONS

The advantages of optical communications are high data rates and low probability of intercept because of narrow beamwidth. These make optical communications a desirable medium for a variety of military applications. One of the functions of greatest potential, however, is for supporting high data rate communications with submerged submarines. There is a window in the blue-green section of the spectrum through which light penetrates sea water with relatively little attenuation. The depth at which the wave's intensity is reduced from its surface intensity by a factor of $\frac{1}{e}$ (where $e = 2.72$) is called the "e-folding" depth of the wave. A submarine may receive a radio wave at a depth of a few times the e-folding depth, depending upon signal intensity at the surface and the receiver sensitivity. Generally, the e-folding depth of a radio wave increases as the square root of the wavelength, so a very long wavelength may penetrate several meters into the ocean before being reduced in intensity by a factor of $\frac{1}{e}$.

ELF radio waves have an e-folding depth of up to 14 meters, blue-green light varies from less than 10 meters to greater than 50 meters, depending on the clarity of the local seawater. The clarity of seawater varies from place to place but does not change very much over time. Seawater may be classified into "ocean optical types," based upon its clarity. Type I is the clearest and Type III is the most turbid. As a general rule, deep ocean water is clear, whereas polar water and some coastal water is turbid. Light with a wavelength of 475 nanometers has the greatest e-folding depth (56 meters in Type I and 10 meters in Type III). Several possible approaches are being investigated for communicating with submarines via blue-green lasers. One approach for one-way communication from shore to submarine involves aiming a land-based transmitter-pulsed beam at a relay satellite that reflects the beam to a submarine's location in a distant ocean. The beam would enter the water with a reduced energy and a cross-sectional area of about 10 square kilometers. At a depth of several e-folding depths, the intensity of the laser beam will be very low, so a very sensitive optical sensor is required. The detector must also filter out sunlight at all but the laser beam's wavelength. This wavelength-selective sensor must be sensitive to light propagating at various angles from the surface.

One such sensor is called the Quantum-Limited Optical Resource Detector (QLORD). This device uses a photomultiplier tube to detect photons and a special filter consisting of a layer of cesium vapor sandwiched between a blue filter on the outside and a red filter on the inside. The way it works is that blue-green light (456 or 459 nanometers) passes through the outer blue filter and is absorbed by the cesium vapor. When cesium vapor atoms absorb light at this wavelength, they fluoresce at infrared wavelengths. The infrared radiation then passes through the inner red filter and activates the photomultiplier tube. Long wavelength sunlight will pass through the blue filter and most wavelengths will not be absorbed by the cesium vapor and will be absorbed by the inner red filter.

The QLORD sensors could be mounted on the hull of a submarine and submarines could receive laser signals at a depth of about 700 meters in Type I water on a clear day or night and about 570 meters on a cloudy day or night. For submarines cruising at any speed, these depths compare favorably with antenna depths at which VLF and ELF signals can be received.

A more advanced design could use a very high power laser on a satellite to broadcast to all locations simultaneously. In both approaches, technology is stressed by the power and reliability required by the laser. Where the laser is on-board the satellite, the reliability issue becomes even more important.

One problem is the need for the transmitter to know the location of the submarine so that it can aim its laser beam. Alternatively, the laser could scan across a suspected area or several possible locations. This procedure would not reveal the submarine's location, but would increase the time required for a submarine to receive a message.

8.5 FIBER OPTICS

The wide bandwidth and low noise of optical fibers make them an ideal transmission medium. Because they are inexpensive and difficult to tap (special fibers are available that make tapping impossible without setting off an alarm), having nonradiating properties, optical fibers are especially suitable for secure communications.

Using optical fiber systems in tactical operations supports force dispersions and enables high-value targets to remain hidden. Optical fibers can be extended 10–20 km or longer without the need for repeaters, whereas the maximum distance for copper cables is only several hundred meters. In addition, optical fibers can be manufactured, even with armor, at much smaller diameters than copper wire (multimode fiber cables have the same capacity as copper cable 27 times thicker). Consequently, they are more compact and can be stored and carried more easily and inexpensively.

Optical fibers are made from abundant and inexpensive material, silica. There are two types of optical waveguide fibers: multimode and single-mode. Good quality multimode fibers can transmit 1000 Mb/s over 1 km. At the lower rate of 100 Mb/s, the repeater spacing can be greater, about 10–20 km.

Single-mode optical fiber systems are designed to propagate light rays along a single direct path along the fiber core, in contrast to multimode systems where the light rays bounce back and forth on many zig-zag paths as they move along the fiber. Without elaborate multiplexing schemes, single-mode optical fiber systems will routinely be able to transmit, without regeneration by a repeater, up to 200 Mb/sec over 80–100 km. In the laboratory, this performance has been exceeded by about ten times.

The major tradeoff in a single mode system is between length (without repeaters) and information rate. The transmission distance of fibers carrying light at 1.3 micrometer wavelength is limited to about 155 km by intense scattering. This decreases to about 80 km as the bit rate is increased, because receiver sensitivity falls off. Chromatic dispersion limits the distance because of the spectral bandwidth of the light source (about 4–6 nm) and because the source's center wavelength may deviate from the fiber's zero-dispersion wavelength.

Interconnection, or splice, loss is the dominant loss factor to be taken into account when a system is designed. In addition to intrinsic scattering, 0.2–0.3 dB/km of extra loss must be allowed for future splicing. There is also a loss when the fiber is bent.

For single-mode fiber systems, the transmitter light source most certainly will be a semiconductor injection laser, since light-emitting diodes, with their relatively low directionality, have a very low input coupling efficiency. They cannot inject enough energy into a single-mode core. There are four basic laser issues that relate to single-mode systems: reliability, temperature dependence, spectral width, and frequency stability. They all have an impact on the performance of the laser as an acceptable transmitter light source. Reliability affects the maintainability of the system. Temperature dependence, spectral width, and frequency stability affect the efficiency and performance of the laser/fiber coupling.

Fibers with comparatively simple step-index and dispersive grading profiles will see use over the next few years. These will provide manufacturing and applications experience and will help to quantify the multi-mode versus single-mode tradeoffs. Alternatively, advanced single-mode fiber designs with improved wavelength-dependent characteristics, minimal dispersion, will become increasingly commonplace in the future.

8.6   NEUTRINO COMMUNICATIONS

Neutrino beams can travel all the way through the earth at close to the speed of light with very little loss. Because of this property, neutrinos, very small electrically neutral elementary particles, have been proposed and studied as a means of communication for submarines. Large arrays of special detectors could sense neutrinos by noting their interactions with other molecules. This technique would work best at great depths where there is little interference from sunlight or cosmic rays. However, the size of the arrays required would be enormous (more than a million cubic meters), the neutrino beam generators would be very large and extremely expensive, and the aiming problem discussed earlier with respect to the blue-green laser would be even more severe for the less effective neutrino detectors.

8.7 LOCAL AREA NETWORKS [3,4]

Local area networks have the following important characteristics: high data rates
(.1–100 Mbps), short distances (.1–50 km), and low error rate (10^{-8}–10^{-11}).
A number of significant benefits can be associated with implementing a local
network. First, and most important, a local network supports system evolution;
additions and replacements can be made with little impact on the other devices
on the network. Therefore, a system can evolve gradually rather than through a
few major upgrades. Other benefits include higher system availability (critical
resources can be duplicated and functions shifted from failed processors to alternate processors with no interconnect problems), sharing of expensive resources, reduction of amount of cabling and interconnections required, easier
integration of equipment and functions, and greater flexibility of equipment
location. Of course, local networking does not guarantee interoperability. The
network only provides connection compatibility, electrical connectivity and perhaps some lower-level protocols. Interoperability requires, in addition, application compatibility, which is not automatically provided by the network.

8.7.1 Transmission Media

The various types of transmission media and their capacities are listed below:

Media	Typical Capacity
Twisted Pair	10 Mbps
Coaxial Cable (Baseband)	10–50 Mbps
CATV (Broadband)	10–20 Mbps Per Data Path
Microwave	1.5 Mbps
Laser	1.5 Mbps
Infrared	250 Kbps
Fiber Optics	250 Mbps
FM Packet Radio	19.2 Kbps

Twisted pair wiring is relatively low cost and typically is preinstalled in many
facilities. It is relatively low speed.

Two types of transmission methods, baseband and broadband, can be employed on a coaxial cable. Baseband systems are typically in the 1–10 Mbps
range. However, by limiting the distance covered and the number of devices
attached, data rates of 50 Mbps can be achieved.

Baseband is a digital signaling technique: the 1's and 0's are transmitted as
discrete voltage processes on the cable. Two limitations are associated with
baseband. First, baseband is applicable only to a bus topology, not a tree. To
achieve a tree, segments of the bus have to be connected via repeaters. Second,
only one signal can be transmitted at a time; the digital signals require the entire
bandwidth of the cable.

Broadband systems are also typically in the range of 1–10 Mbps for a single data path (e.g., frequency channel), with 20 Mbps representing a practical upper limit. However, broadband can support multiple-data paths. Broadband systems use CATV (community antenna television) cable. With a broadband cable, the frequency bandwidth of the cable (350–400 MHz) is divided into channels. Data is modulated onto a carrier frequency, using the center frequency of one of the channels.

Broadband is inherently unidirectional. To achieve full connectivity, two different techniques are used; mid-split and dual. Mid-split accomplishes two-way transmission by the use of bi-directional amplifiers, whose filters allow approximately half of the available spectrum to be transmitted in each direction. The cable is terminated at a point called the head-end. Inbound transmission, toward the head-end typically uses 5-116 MHz; outbound transmission uses 168-300 MHz. The head-end contains a device for translating inbound frequencies to outbound frequencies.

The dual cable system consists of two unidirectional cables joined by a passive head-end. One cable is used for inbound transmission and the other for outbound. The head-end passes signals across the entire spectrum. Stations send and receive on the same frequency.

There are three kinds of data transfer: dedicated, switched, and multiple access. For dedicated service, a small portion of the cable's bandwidth is reserved for exclusive use by two devices. No special protocol is needed. Transfer rates can be anything up to 20 Mbps.

The switched technique requires setting aside a number of frequency bands. Devices are attached through "frequency agile" modems. Initially, all attached devices, together with a controller, are tuned to the same frequency. A station wishing a connection sends a request to the controller, which assigns an available frequency to the two devices. Transfer rates would typically be 64 kbps or less.

The multiple access technique is the one most commonly used. To control this technique, a multiple access protocol is needed. Broadband multiple access systems use the same techniques and protocols that are employed for baseband. They are discussed more fully in Section 8.7.3.

Fiber optic cables can support extremely high data rates, up to 250 Mbps. Fiber optics have many other advantages, as we have noted earlier.

Another possible candidate is FM packet radio. FM radio is a broadcast rather than a directional point-to-point technique. Therefore, it is less vulnerable to physical obstruction. It is also relatively immune to atmospheric problems. Because it is broadcast, nodes on the network do not have to be stationary; mobile FM transceivers can be used. The components are relatively cheap. Its principal drawback is its comparatively low data rate of 19.2 kbps.

8.7.2 Topology

Local area networks are frequently characterized in terms of their topology. Three topologies are common: star, ring, and bus or tree.

In a star topology, a central switching element is used to connect all the nodes in the network. The central element can be a message, circuit, or packet switch.

The ring topology consists of a closed loop, with each node attached to a repeating element. Data circulates around the ring on a series of point-to-point data links. A station needing to transmit waits its turn and then sends its data in the form of a packet. The packet contains data plus source and destination address fields. As the packet circulates, the destination node copies the packet data into a local buffer. The packet continues to circulate, returning to the source node as a form of acknowledgment. Because the ring is constructed as a series of point-to-point links, almost any transmission media can be used. The ring is often the most efficient topology for computer-to-computer communications (particularly under heavy loading).

The most common type of local area network in use today is a bus or tree topology using copper cable. The bus or tree topology is characterized by the use of multiple access, broadcast media. The bus is a special case of the tree; it is a tree with a single trunk and no branches. Because all devices share a common medium, only one pair can communicate at a time. A distributed medium access protocol is used to determine which station may transmit next. As with the ring, transmission employs a packet containing source and destination address fields. Each station monitors the medium and copies packets addressed to it.

Tree topology networks typically use copper cable. Twisted-pair is usually too low a capacity. Fiber optic cable is not currently practical as a multi-access medium because multi-drop attachments are not yet commercially feasible.

8.7.3 Protocols

Two protocols have been developed and are being standardized for controlling access to a broadcast, multi-access bus: carrier sense multiple access with collision detection (CSMA/CD) and token bus.

The CSMA/CD technique works as follows. Traffic already on the network has priority. If one station is transmitting, all the other stations listen. Stations on the bus determine if the channel is busy by listening for the carrier frequency signal (broadband) or a voltage pulse train (baseband). When the channel is clear, any station may transmit. If two stations attempt to transmit at the same time, they will detect a garble or collision. The affected stations will then cut short their transmission and wait a period of time before attempting retransmission.

In the token bus technique, the stations on the bus form a vertical ring. They are assigned positions in an ordered sequence, with the last member of the sequence followed by the first. At start-up time, one station is granted permission to transmit. It sends packets and waits for acknowledgment. All other stations listen. They may secure data packets and acknowledge them. When the first station is finished, it transmits a packet containing a control pattern known as a token to the next station on the vertical ring. This station then has permission to transmit, and will pass on the token upon completion.

In an actual ring topology, the token ring protocol provides efficient utilization of the ring topology. It works much the same as the token bus technique just described, but the station logic is simpler. When a station wishes to transmit, it monitors the ring until it detects the token pattern. The station removes the token from the ring and transmits a frame of data. When the frame has circulated and returned, the initiator places a token back on the line. The token will circulate until it encounters a station with data to transmit. One way that token bus differs from token ring is that the token bus passes the token to the device with the next address no matter where it is physically located. For token ring, the token is passed to its physical neighbor.

8.7.4   LAN Evolution

Initially, the concept of local area networking became familiar as a data communication scheme dependent on private lines, public switched service, and private switched systems accommodating specific systems and application environments. The local networking concept evolved to include a general-purpose, multivendor/system environment that provides interconnection of a variety of terminals and computers within one building or in several buildings in close physical proximity. Modern local networking systems are usually characterized by bandwidths of consecutive frequencies and high data rates of several million bits per second. These systems use coaxial cable, twisted-pair wire, or fiber optic transmission media. Baseband/broadband LAN development is now focusing on the standards developed by the IEEE-802 Standards Committee. The ISO Committee on Open Systems Interconnection (OSI) incorporated the IEEE-802 Committee LAN standards into the OSI model.

In the past few years, a subset of LAN systems has evolved—called the Personal Computer Network (PCN). The PCN is a natural outgrowth of the automated office environment dependent on desktop personal computers for data and word processing applications. Therefore, personal computer users requiring low-to-moderate performance are looking to PCNs as a low-cost but effective networking solution. Many PCN configurations, for example, require support for no more than six personal computers. The key requirement for PCNs is resource sharing of letter-quality printers and high-capacity disk systems. File/print and communication servers can provide shared access for personal computers/workstations in a PCN environment. Electronic Mail and Electronic Files are often enhanced application features. Usually, twisted-pair or coaxial cable supports up to 1 Mbps data rates (some are 2.5 Mbps or higher), and the typical price range for a single node-to-network interface is from $200 to $600 per computer connection. Many new entries to this LAN arena are PCNs.

8.7.5   Standards

The IEEE-802 Committee has been the force behind the development of a family of LAN standards. These standards are based upon the ISO OSI reference model, which consists of seven layers. These layers are:

> Layer 1—physical
> Layer 2—link
> Layer 3—network
> Layer 4—transport
> Layer 5—session
> Layer 6—presentation
> Layer 7—application

Standards now exist for physical and link layers as defined by the ISO Open Systems Interconnection (OSI) reference model.

In addition to these standards, there are the IBM SNA standards, which are also based on a layered architecture similar to the OSI model.

The IEEE-802.1 subcommittee is charged with establishing the relationship of IEEE-802 standards and the OSI reference model. This group has incorporated the LAN standard into the OSI model. It now appears that the IEEE standards will be identical to those adopted for the OSI model.

The IEEE-802.2 subcommittee has established a standard to provide a common logical link control protocol.

The IEEE-802.3 proposal is now the standard for a baseband bus LAN using CSMA/CD as the access control method for the physical layer. Four transmission rates are endorsed—1M, 5M, 10M, and 20M—although the adopted standard is for a 10 Mbps transmission rate. The committee is now evaluating a proposal for a 1 mbps version of the 802.3 LAN.

The IEEE-802.4 proposal is now a standard for a baseband or broadband bus LAN using token passing as the access control method for the physical layer. The standard is now being revised to overcome problems implementers have encountered in some of the specifications. The revision is loosening up some of the broadband specifications, changing the differential Manchester coding scheme to straight Manchester encoding, and incorporating some appendix material into the main body of the report. Transmission rates are 1Mbps, 5Mbps, and 10M bps.

The IEEE-802.5 proposal is for a baseband ring LAN using token passing as the access control method for the physical layer. This standard has reached the draft proposal stage. It provides for using 100-ohm twisted-pair cable and data transmission rates of 1Mbps and 4Mbps.

The IEEE-802.6 subcommittee is working on proposals for metropolitan area networks. Proposals have been made for TDMA (time division multiple access), polling, and cellular Ethernet networks. The company that submitted the TDMA proposal went out of business, and no one else has submitted another TDMA proposal. The polling proposal will probably be resubmitted for final approval.

The IEEE has two more subcommittees that operate as technical advisors on broadband LANs (802.7) and fiber optic LANs (802.8) to the other subcommittees.

PCN vendors are digressing from the beaten path to respond to user pressures for speedy solutions to personal computer networking. A standard-for-PCNs movement is surfacing; some of the new PCNs include parts of the OSI guidelines (seven-layer computer-network architecture) in their network systems design in anticipation of a need for standards.

The IEEE-802.3, 802.4, and 802.5 subcommittees are all considering 1M-bps versions of their respective standards. Once VLSI chips are available to implement the protocols, the IEEE-802 LAN standards will be appropriate for implementing PCNs.

Another important recent development has been the increase in the use of fiber optic cable in LANs. Ungermann-Bass now offers a total fiber optic version of its Net/One LAN. The primary use of fiber optic cable, however, is to extend a coaxial cable LAN or to interconnect coaxial cable LANs in separate buildings. A number of fiber optic LANs are in fact installed, but vendors appear reluctant to offer them as standard products.

8.7.6 LAN Specifications

Included on the following pages is a summary of available LANs and their characteristics. The data in Figure 8.2 is reproduced from the 1984 Data Decisions publication [5]. Each entry is initially identified by network name. Specifications are categorized by type of network, special features, transmission speed, maximum cable length, applications, configurations, interfacing, gateways, date of first announcement, number installed, and pricing.

Type identifies the kind of system media (cable, bus, wire) and access method, such as contention and token passing. CSMA/CD or Carrier Sense Multiple Access with Collision Detection is a contention scheme used in Ethernet.

Features that distinguish between network types include the following: "nonblocking," "collision avoidance," and "positive acknowledgment."

Nonblocking is associated primarily with PBX-type local area networks to define the system's traffic-carrying capacity. Usually, a system can support only a fraction of its total user capacity simultaneously. The assumption is made that all users will not want access to the system at the same time and all the time. Nonblocking means the system can support its peak capacity simultaneously and continuously.

Collision avoidance is a scheme to avoid collisions using the CSMA access scheme. For example, on the CORVUS OMINET, a second user waiting to transmit on a busy line is signaled to transmit later at an arbitrary time. On

Figure 8.2 Local Area Network (LAN) performance summary [5].

COMPANY	NETWORK
A.B. Dick	The Loop
Alpa Computer Inc	ALSPA-NET
Altos Computer Sys	WorkNet
AT&T Teletype	Teletype 4840 LC
Amtel Systems	Messenger
Apollo Computer	DOMAIN
Applitek Corp	UniLAN
AST Research	PCnet
AST Research	PCnet II
Bragen Corp	ELAN
Bridge Comm Inc	Ethernet
Codenoll Tech	Ethernet
Codex Corp	Net/One
Complexx Sys, Inc	XLAN
Compucorp	OmegaNet
Computer Network	DLX-10
Computer Network	DLX-320
Concord Data Systems	Token/Net
Contel Info Sys	ConTelNet
Contel Info Sys	STAR-Eleven

Column categories (dot-matrix chart):

- NETWORK TYPE: Broadband, Baseband
- ACCESS METHOD: Token passing, Contention, Other
- TRANSMISSION SPEED: to 10K bps, to 24K bps, to 104K bps, to 1M bps, over 1M bps
- CABLE LENGTH: to 2000 ft, to 5000 ft, over 5000 ft
- GATEWAYS: IBM SNA/SDLC, X.25, Xerox Ethernet, Other
- APPLICATION AREA: General Business, Electronic Mail, Word Processing, Industrial, Other

Figure 8.2 (continued) — Local Area Network (LAN) performance summary [5]

Column categories (left to right): NETWORK TYPE (Baseband, Broadband); ACCESS METHOD (Token passing, Contention, Other); TRANSMISSION SPEED (to 1M bps, to 2M bps, over 10K bps, over 10M bps); CABLE LENGTH (to 2000 ft, to 5000 ft, over 5000 ft); GATEWAYS (IBM SNA/SDLC, X.25, Xerox Ethernet, Other); APPLICATION (General Business, Electronic Mail, Word Processing, Industrial, Other).

COMPANY	NETWORK
Convergent Systems	Local Resource Shar
Corvus	Omninet
Cromemco	C-Net
Data General	XODIAC Network Bus
Datapoint	ARCnet
Davong Systems Inc	MultiLink
DBS International	DBS-NET
DESTEK Group	DESNET
Develcon Elect Inc	Develnet
Digital Equipment	DECdataway
Digital Equipment	Ethernet
Digital Microsystems	HiNet
Doelz Network Inc	Doelz Network
Gandalf Data	PACXNET
Gateway Comm, Inc	G/Net
General Electric	GEnet
Gould	MODBUS
Gould	MODWAY
Hewlett-Packard	Interface Bus
Hewlett-Packard	LAN 9000
Hewlett-Packard	SRM
Honeywell	TDC 3000
Iconix Corp	Cinchnet

Figure 8.2   Local Area Network (LAN) performance summary [5] (continued).

Company	Product
Inforex, Inc	ULTRANET
Intecom	LANmark
Interactive Sys/3M	VIDEODATA
Interactive Sys/3M	LAN/1
Interlan	NET/PLUS
Int'l Bus Mach (IBM)	8100 Loop
Int'l Bus Mach (IBM)	Series/1 Ring
Int'l Bus Mach (IBM)	PC Cluster
Intertec Data Systems	CompuStar
Lanier	LBS 5000
Logica	Polynet
M/A-COM DCC	Infobus
M/A-COM Linkabit	IDX-3000
Micom Systems	INSTANET
Molecular Company	SuperMicro Multiuser
Morrow	MORROW NETWORK
NBI, Inc	NBINET
NCR Corp	Decision Net
NCR Corp	MIRLAN
Nestar Systems	PLAN 4000 Series
Network Systems	HYPERchannel
Network Systems	HYPERbus
North Star Computer	North Net
Novell Data Systems	ShareNet
Orange Compuco	ULCnet
Orchid Technology	PCnet/PCnet Plus
Percom Data Corp	PerComNet

Figure 8.2 (Continued).

COMPANY	NETWORK	NETWORK TYPE: Baseband	NETWORK TYPE: Broadband	ACCESS METHOD: Contention	ACCESS METHOD: Token Passing	ACCESS METHOD: Other	TRANSMISSION SPEED: to 1M bps	TRANSMISSION SPEED: to 2M bps	TRANSMISSION SPEED: to 10M bps	TRANSMISSION SPEED: over 10M bps	CABLE LENGTH: to 2000 ft	CABLE LENGTH: to 5000 ft	CABLE LENGTH: over 5000 ft	GATEWAYS: IBM SNA/SDLC	GATEWAYS: X.25	GATEWAYS: Xerox Ethernet	GATEWAYS: Other	APPLICATION AREA: General Business	APPLICATION AREA: Electronic Mail	APPLICATION AREA: Word Processing	APPLICATION AREA: Industrial	APPLICATION AREA: Other
Perq Systems	Ethernet	●	—	●	—	—	—	—	●	—	—	—	●	—	—	—	—	●	●	●	—	●
Pragmatronics	TIENET	●	—	●	—	—	●	—	—	—	—	—	●	—	—	—	●	●	—	—	—	●
Prime Computer	RINGNET	●	—	—	●	—	—	—	●	—	●	—	—	—	●	—	●	—	—	—	●	●
Prolink	Proloop	●	—	●	—	—	—	—	●	—	●	—	—	—	—	—	—	●	—	●	—	●
Proteon	proNET	●	—	—	●	—	—	—	●	●	—	—	●	●	●	●	●	●	●	●	●	●
Racal-Milgo	PLANET	●	—	—	●	—	—	—	—	—	—	—	—	—	—	—	—	—	—	—	—	—
Santa Clara Systems	PCnet	●	—	●	—	—	●	—	—	—	—	—	●	—	—	—	—	●	●	—	—	—
Scientific Data	SDSNET	●	—	●	—	—	●	—	—	—	—	●	—	—	—	—	—	●	●	●	—	●
Sidereal Corp	MIC-LINK	●	—	●	—	—	●	—	—	—	●	—	—	—	—	—	—	●	●	—	—	●
Sperry	SHINPADS	●	—	—	—	●	—	—	●	—	●	—	—	—	—	—	●	—	—	—	—	●
Starnet Data Systems	Starnet II	●	—	—	—	●	●	●	●	●	—	—	●	●	—	—	●	●	●	●	●	●
Stratus Computer	StrataLINK	●	—	●	—	—	●	●	—	—	—	—	●	●	●	—	—	●	—	●	●	●
Syntech	MARS/NET	●	—	●	—	—	●	—	—	—	—	—	●	—	—	—	—	●	—	●	—	●
Syntrex	SYNNet	●	—	●	—	—	—	—	●	—	—	●	—	—	—	—	—	●	—	—	—	—
Sytek	LocalNet 20	—	●	●	—	—	●	—	—	—	—	—	●	—	●	●	●	●	—	—	—	●
Sytek	LocalNet 40	—	●	●	—	—	—	—	●	—	—	—	●	—	●	—	●	●	—	—	—	●
Tandy Corp	ARCnet	●	—	—	●	—	—	—	●	—	—	—	●	—	—	—	—	●	—	●	—	●
Teltone	DCS-2/2S Data Carrier	●	—	—	—	●	—	●	—	—	—	—	●	—	—	—	—	—	—	—	—	●
3COM	Ethernet/UNET	●	—	●	—	—	—	—	●	—	—	—	●	—	—	●	—	●	●	—	—	●
3COM	Etherlink	●	—	●	—	—	—	—	●	—	●	●	●	—	—	—	—	●	●	—	—	●

Figure 8.2 Local Area Network (LAN) performance summary [5] (continued).

Ungermann-Bass	Net/One Baseband
Ungermann-Bass	Net/One Broadband
Ungermann-Bass	Fiber Optic Net/One
Ungermann-Bass	Net/One Thin Coax
Valmet	Millway
Vector Graphics	LINC
Wang Laboratories	Wangnet
Xerox	Ethernet
Xyplex, Inc	XYPLEX System
Zilog	UNET
Ztel	AXIS

Figure 8.2 (continued).

Network Systems HYPERbus, collisions are avoided by using a "virtual" token passing scheme for line access.

Positive Acknowledgment is a scheme in which a user must obtain acknowledgment before transmission can take place.

Transmission Speed indicates speed in millions (M) of bits per second (bps).

Cable Length describes the maximum length of cable from end-to-end, node-to-node, device-to-device; also includes aggregate cable length using repeater or extender.

Application describes functional use such as office, industrial, or financial.

Interfacing includes required device, adapter, or standard that devices must use to connect to the trunk or cable used in the local network.

Gateways are provisions for crossing the boundary from local network into another type of local or distributed network such as Ethernet, SNA, X.25 networks, or other local nets.

First Announced refers to year of announcement, if known.

Number Installed is as of publication, if known.

Pricing includes purchase price of major network components.

8.8   OTHER TECHNOLOGY OPPORTUNITIES

Listed below are other technology opportunities that can be exploited to provide a more effective C^3I system. More details can be found in the cited references.

- Microelectronics Technology—Included in this category are charge-coupled devices, surface-acoustic-wave devices, and technology advances in VLSI/VHSIC (very high speed integrated circuits) [6,7,8,9].

- Microprocessor Advances—Distributed processing has been enhanced by these advances [8,10].

- Adaptive Arrays—These arrays can automatically form and steer beams (including tropoapplications) and nulls (against jammers) [11,12].

- Multiple Beam Antennas—Potentially, these antennas, with agile switching, could be used for both communications and radar applications, and combinations thereof [13,14].

- SHF/EHF Technology [15,16]. Transmission at these high frequencies allows for greater data rates, more directivity, and greater resistance to jamming. At the higher EHF frequencies, receive antennas can be quite light and agile.

- Onboard Processing for Satellites and Airborne Relays—adaptive circuitry and ultrahigh reliability are important considerations here [3].

- Conformal Arrays—Ideally, conformal arrays offer the promise of excellent aircraft antenna capabilities with little or no modification to the airframe and minimum aerodynamic effects [2,17].

- Narrowband Voice Digitization—Included in this category are limited-vocabulary, word-recognition, and word-synthesis systems, which can reduce bandwidth requirements significantly and yet handle many tactical emergency messages [18].

- Advanced Memory Technology—Inexpensive random-access memory technology may reduce significantly the software problems that have plagued many development programs by allowing programmers more freedom to structure their programs.

- Cryptography Advances (including Public Key Cryptosystems) allow for secure transmission of information through encoding and scrabbling [19,20].

- Fault-Tolerant Processors and Systems permit the design and implementation of highly available and reliable systems that can continue to operate in the presence of system component failures [14,15].

- Encoder/Decoder Technology (in particular, advanced jam-resistant and error-correction technology)

- Nuclear-Resistant Devices and Circuits—Optical devices and circuits such as light valves and optical transmission channels are particularly promising.

- Broadcast Channels and Multiway Channels—Examples of "broadcast channels" and "multiway channels" are presented in [4], which also indicates their applicability to some important communications problems.

- Display Technology—Display of information is an important factor in communications system effectiveness for several reasons: to aid the network-control personnel, to help personnel at the information source (e.g., to determine what information really should be transmitted, to display the received information, and to aid in equipment monitoring and fault isolation. Display technology is closely coupled to the science of decision aids [21].

The parameters for many of the technologies discussed above are currently being determined by commercial applications, not military applications. The potential for these technological opportunities will be discussed in the subsections that follow.

8.9 SYSTEM APPLICATIONS

Tactical communications for the 1990s will be improved not only by exploiting the technology opportunities identified previously, but also by "system applications." Sound system applications can yield more cost-effective communications both by using the new technology advances in a cost-effective manner and by employing the equipment currently being acquired in a better manner.

A fundamental part of the process of identifying and evaluating promising system-oriented approaches and applications is a strong system-planning activity. Unless an appropriate system "architecture" is established and a strong system-engineering/analysis effort is conducted, new technology advances and equipment acquisitions will not be employed in the most cost-effective manner. The TAFIIS Master Plan, which provides the overall C^3 context for the Air Force's tactical communications-planning/architecture effort, has spawned several subordinate communications-planning activities. For example, the Air Force conducted an intensive study of possible solutions to ground-to-ground communications problems of the Tactical Air Forces for the 1990–2000 era. Similar recommendations are being made for the Army as a result of the Army C^2 Master Plan (AC^2MP) and for the Navy as a result of the Navy C^2 plan.

General guidelines for eliminating or reducing the "remaining problem areas" through sound system design are as follows:

- Reduce unintended radiation in unintended directions; concomitantly, reduce reception of undesired radiation from unintended sources and directions, and via unintended ports.
- Make the friendly signals and sources less identifiable and less detectable by burying them in the natural background as much as possible, by putting them in a frequency region the enemy is not capable of detecting (because of the propagation characteristics), or, if possible, by not radiating them.
- Make the communications system more distributed and redundant, with fewer critical nodes and more backup links.
- Deny the enemy the capability to attack communications nodes by denying him location information, spreading out the source so it is not an attackable target, deceiving him with false or disguised signals and targets, or turning off the source at critical times.
- Make the communications system rapidly reconstitutable when parts are destroyed.
- Secure the system further by making more of the transmitted signals as undecipherable as possible, making access to the system impossible, and making the system immune to signal penetration.
- Provide more capability to discriminate between undesired and desired signals, by frequency, time, waveform, direction of arrival, etc.

An important constraint that must be considered in all evaluations of possible system changes is "backward compatibility"; that is, a major part of the tactical systems cannot be shut down totally while new equipment is introduced and debugged.

Some specific promising system applications are identified and discussed below.

8.9.1 Reduced-Information-Rate Operation

The intense electronic warfare threat that is anticipated in the 1990s, and its drastic impact on communications equipment, indicates that every effort should be made to identify equipment and operational techniques that will allow significant reductions in the amount of information that must be transferred. In many battlefield situations—both airborne and groundbased—it will not be possible to transmit the wide bandwidths that would be most desirable for many military operations (e.g., voice, imagery). Techniques to significantly reduce the bandwidth requirements associated with voice and imagery information, and then use spread spectrum to spread the resulting bit stream over a wide bandwidth for AJ purposes, must be investigated more thoroughly to determine their proper application. Such techniques should include the substitution of nonvoice messages for voice interchanges and the use of limited-vocabulary voice. An important aspect of the development cycle leading to the fielding of equipment based on such techniques is extensive experimentation in operational situations; optimal substitution of digital data for voice, for example, will require extensive man-machine experimentation. Some user dissatisfaction is to be expected, but in some battlefield situations, the basic information-transfer issue may ultimately resolve to "Do you want 75 bits/s or nothing?"

8.9.2   Network of Ground Relays

Line-of-sight links and pseudo-LOS links (e.g., diffraction paths) show great promise for providing the requisite combination of jam resistance and mobility needed in the battlefield environment of the 1990s. A network of ground relays mounted on mobile military vehicles could implement such links. Adaptive-array antennas and straightforward LOS antennas should both be considered [1].

To illustrate the netting approach, Figure 8.3 depicts a representative network for current tactical ground-to-ground communications overlayed with a possible new network of ground relays. The current troposcatter network (indicated by double solid lines) is representative of that for the Tactical Air Control System;

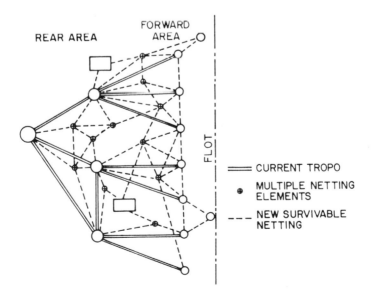

Figure 8.3   Representative survivable network of ground relays.

the sources and sinks of information are, in general, ground-based radars and control centers. (Two ground-based forward air controllers are included in the figure, even though they do not use troposcatter links, because they might use the new netting approach.) A new network, much more survivable than the current connectivity, could be developed by adding new nodes, indicated by $+$, and adding the dashed links. Another, more advanced version of this network can be conjectured, wherein a multiplicity of radar types (monostatic ground-based radars, airborne radars, ground-mobile radars, multistatic radars, etc.) are interconnected. A natural adjunct to such a network would be highly mobile, combined radar/communications equipment such as that depicted in Figure 8.4. The basis for this concept is a circular phased-array radar that could "pop up," perform a limited number of search scans, go down, and move on. The data communications would be effected by communications beams using the same antenna and transmitter. This could be accomplished by searching for LPI monitor signals from friendly communications partners, rapidly locking onto these signals, and communicating in short bursts. The phased array provides an adaptive antenna-nulling capability as well. The LPI and mobility features would provide excellent survivability.

The cost of the individual relay elements, and foliage-induced propagation losses, will be key factors in determining the cost-effectiveness of this approach.

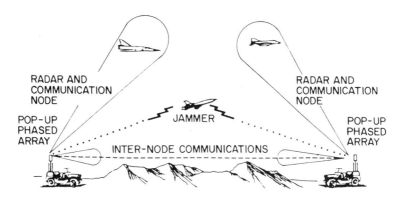

Figure 8.4 Mobile communications/radar concept [22].

8.9.3 Packet Switching with Adaptive Routing

Packet switching breaks up long digital messages into short blocks called "packets." Each packet is provided with address information and is passed from switch to switch, briefly stored at each switch, until it reaches its final destination. There the packets, which may arrive via different paths, are assembled into the original long message. (See [23] and [24] for basic information concerning packet switching and its evolution.) Each switch maintains up-to-date information concerning the available paths through the network, which usually is highly interconnected; this information allows adaptive selection of routes through the network.

The adaptive routing can be an extremely important countermeasure against high-powered airborne jammers that are attempting to jam large portions of the terrestrial network. As such jammers fly along, their geometric relationship to the receiving antenna changes significantly, and terrain effects can change air-to-ground propagation losses tremendously. These factors result in links being jammed, then reestablished, then jammed, and so on. Dynamic, adaptive routing can effectively find paths through such a changing network, storing some information for later transmission if no path is currently available. Of course, dynamic, adaptive system control is a prerequisite for obtaining the full benefits of this potentially powerful system approach.

8.9.4   Packetized Voice/Data Network

Initially, packet switching was employed only for nonvoice messages. Now a great deal of research is being conducted into the possibility of breaking digitized voice streams into packets and handling them in the same manner as the data packets. (See, for example, Coviello's comparison of packet-switched voice and circuit-switched voice for various network architectures and voice digitization rates [25].) With adaptive routing, this approach could provide significantly improved survivability in a battlefield environment. The question of the cost effectiveness of the packetized voice/data combination is being addressed in ongoing Air Force planning efforts

8.9.5   Packet Radio

Packet radio [26] extends the concept of packet switching to the domain of multiple-access radio channels. It can satisfy the needs of large numbers of mobile subscribers requiring information distribution over a wide geographic area (e.g., an Army division). It is especially efficient if the information is "bursty," as in computer-to-computer communications. Each packet radio contains a microprocessor that breaks up the messages to be transmitted into packets (as in packet switching) and provides dynamic control over the radio's access to the channel to minimize conflicts (overlapping transmissions) between radios. Effective spectrum utilization and cost will be key factors in determining the utilization of packet-radio-based networks for information distribution.

Desired characteristics of a packet radio network are: transparency (no special user actions required); area coverage and connectivity; mobility and rapid deployment; interoperability among nets; throughput with low delay; control, routing, and addressing options; and various application-directed services. As is the case for packet switching, a packet radio network could be designed to accommodate real-time speech and end-to-end encryption. In order for the equipment to resist electronic countermeasures (ECM), provide LPI, and perform well in the natural environment, spread-spectrum techniques (pseudonoise modulation and/or frequency hopping) would be used. This can lead to a secure, integrated communications, navigation, and identification capability.

8.9.6   JMTSS

The Joint Multichannel Trunking and Switching System is an important concept that, if properly designed, could help solve some of the Air Force's tactical ground-to-ground communications problems. Thus far, an architecture and implementation for the JMTSS have not been selected; however, the *basic concept* of having joint-service communications facilities available in a battlefield environment is very important. Under control of the theater commander, such facilities could provide access points for all of the services, and common-user trunking and switching in the battlefield. Potentially, a joint-service network could provide redundant, jam-resistant, survivable communications facilities for ground-to-ground messages. Depending upon the architecture that evolves from ongoing investigations, rapid setup times could also be achieved for communications via an LOS link to the nearest access point. The adaptive relays described in [27] are promising candidates to effect such links, but standard LOS approaches may also be cost-effective in some situations.

8.9.7   Airborne Relays

Airborne relays hold significant promise for improving the tactical communications in the 1990s. A high-altitude airborne relay could provide extensive capacity for ground-to-ground communications, perhaps as a backup for a satellite that has been jammed or destroyed. (A large ground-based "sanctuary jammer" located in enemy territory far from the battle area could not affect the airborne relay.) A smaller, lower-flying airborne relay could "patch" a portion of a ground-based network that has become disconnected because of enemy action. A properly positioned airborne relay could effect a connection between two aircraft that are not within line-of-sight of each other, and still be beyond LOS of ground-based jammers.

Many promising candidates exist for tactical airborne relays: remotely piloted vehicles (RPVs), truck-launched vehicles under development by the Air Force (e.g., the LOCUST) and the Army (e.g., the AQUILA), tethered aerostat balloons, etc. It would certainly be best if all of the services employed the same or compatible airborne relay(s).

One example of the use of airborne relays is an emergency application in which an RPV would fly a well-controlled "racetrack" orbit at an altitude of approximately 60,000 ft. In a tactical situation, 30 terminals might be in the line-of-sight coverage area below the RPV. UHF, SHF, and low EHF frequencies have been considered for this application, with a TDMA mode of operation.

8.9.8   Improved System Control

The initial versions of TRI-TAC's Communications System Control Element will provide only those network-control functions that have been identified as "minimum essential." Other functions and capabilities should be added in an evolutionary manner, and the definition of these functions and capabilities should be given high priority in the evolution of the CSCE. Some candidates are as follows:

- Calculation of routing-table changes on the basis of real-time traffic flow information
- Automated capability for terrestrial link design
- Calculations related to the impact of satellite-channel reallocations on the reconstitution of a damaged ground-to-ground network

Additional man-machine experimentation is necessary in the near future to identify the major improvements that should be made to the CSCE software and hardware for future versions. Such improvements will provide increased survivability for the Air Force's ground-to-ground communications by dynamically rerouting essential traffic around jammed links and damaged nodes.

8.9.9 Distributed System Control

Beyond improvements to the CSCE within the basic TRI-TAC architecture, distributing the system-control functions widely throughout the theater holds promise for improving survivability against direct attacks on the CSCEs. Major advances in microprocessors suggest that extensive system-control capabilities could be installed at many locations. Dispersing network-control functions (especially those not requiring manual intervention) would allow the network to continue operation even after many of its nodal points have been destroyed.

8.9.10 Better Routing Procedures

Even without the introduction of major changes in the currently planned TRI-TAC switches, additional survivability could be obtained by modifying the TRI-TAC routing plans to make them more flexible and more adaptive to a dynamically changing tactical network. Increasing the number of alternate routes available at the originating switch, and selecting alternate routes that are not all dependent on one or two critical links, would improve network performance in a battlefield environment. These routing-plan changes will become more viable and more advantageous when the TRI-TAC-developed equipments have become the predominant equipments in the field, and when higher connectivity among elements becomes available. More-adaptive routing procedures also make the network less vulnerable to the loss of CSCEs.

8.9.11 Modular C³ Concept

Potentially, standardized C^3 modules (both hardware and software) can reduce the cost of tactical C^3 systems and improve their survivability by distributing important functions throughout the battlefield or by allowing such functions to be reassigned when centers/nodes are destroyed or communications links are jammed. As a step toward such standardized modules, a modular C^3 concept is being developed and investigated. As indicated in Figure 8.5, the basic concept is to develop a set of interfaces for each module in a C^3 center so that all modules would communicate over a common "bus" [4]. Additional modules could be added via bus interface units, allowing the center to expand, contract, or be reconfigured. These standardized modules could then be used as "building blocks" for a variety of C^3 centers and their communications elements. Only a small number of modules would be unique to a particular kind of center. Fiber optics technology is an important candidate for implementing the bus within the center. As the transition from analog to digital communications progresses in the 1980s, the opportunities for employing modular C^3 concepts will increase. A more efficient, essentially all-digital, communications architecture will probably be found to be more cost effective, and the hybrid analog/digital architecture and equipment currently being developed under the TRI-TAC program will eventually be superseded by all-digital elements.

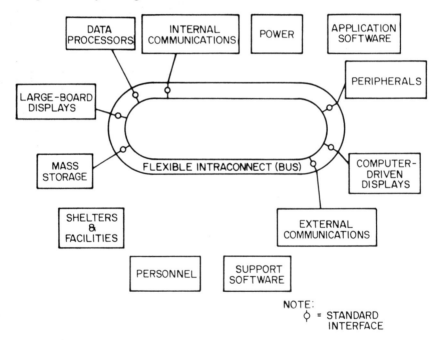

Figure 8.5 Modular C^3 concept.

8.9.12 Distributed Databases

Distributing vital databases concerned with ground-to-ground communications throughout many battlefield locations will improve the survivability of the network. Tradeoff analyses concerning the degree to which the databases should be distributed and replicated are currently being conducted.

8.9.13 Multiple Relays

A key factor in determining the signal-to-jamming power ratio is the relationship between the friendly transmitter-receiver distance and the jammer-receiver distance. The use of multiple relays—airborne or ground-based—can significantly reduce the friendly transmitter-receiver distance and transform a jammed link into an operational link. Cook [28] provides a quantitative analysis of the advantages to be gained by this approach. Key factors in the evolution of an efficient implementation of the JTIDS concept will be careful consideration of the proper use of airborne relays, and net management in general.

8.9.14 High-AJ Satellite

The use of a high-technology satellite that is tailored to provide communications services for a combat theater may represent important communications capability in the electronic-warfare environment of the 1990s. Considerable promise is offered by the use of: the EHF band, adaptive arrays for nulling jammers, multiple narrowbeams (with onboard switching between beams), onboard processing to obtain ECCM, variable rates on the various beams, onboard storage, and a bus architecture. Intersatellite links could provide connections to out-of-theater facilities. Unfortunately, none of these high-technology approaches reduces the vulnerability of the satellite to direct physical attack.

Cummings et al. [15] have analyzed the potential for future use of EHF in MILSATCOM systems. The advantages they point out are: (1) relief of potential frequency and orbital congestion; (2) improved AJ due to the use of high-resolution antenna nulling, broader bandwidth spread-spectrum modulation, and closer parity with interfering sources; and (3) better LPI characteristics. Important offsetting factors are: (1) cost; (2) the fact that antenna gain is limited by the dimensional accuracy (with respect to the wavelength) achievable in the manufacture of the antennas at these frequencies; (3) tracking accuracy limitations; (4) rain attenuation; and (5) other atmospheric effects. (These factors are less significant at 30 GHz than at the higher EHF frequencies.) In order to minimize path length through the atmosphere and potential rain areas, higher elevation angles for earth terminals are desirable, compared with lower frequencies; this leads to the consideration of orbits other than, or in addition to, geostationary orbits (e.g., elliptical polar orbits) for some applications.

With regard to the satellite, EHF offers the advantage of narrow higher-resolution beams that can be switched to focus on certain spots on the earth and discriminate against others by selective nulling. Furthermore, aircraft terminals can utilize conformal arrays to reduce aerodynamic drag and space/mechanical requirements, as well as to provide adaptive nulling and beam switching. EHF has also been proposed for intersatellite covert links.

The use of laser communications in appropriate regions of the visible, ultraviolet, and infrared spectra is being investigated for ground/satellite uplinks and downlinks and for intersatellite links. The considerations are similar to those for

EHF; however, the constraints on beamwidth and tracking accuracy are even tighter. Propagation through the atmosphere is strongly dependent on clouds and weather conditions.

8.9.15   Anti-ARM Approaches

A system-oriented combination of approaches can significantly reduce the vulnerability of major communications facilities to antiradiation missiles (ARMs) in the 1990s. LPI techniques (e.g., beamshaping, spectrum spreading) can reduce the ability of the enemy to locate the friendly transmitter. Radar (ground-based or airborne) can detect the approach of an airborne threat to the friendly transmitter, and transmit this information to the radiator (possibly via a ground-based decision-making center); the radiator is then turned off and decoys are turned on in the vicinity of the radiator to confuse or misdirect the ARM.

8.9.16   Laser and EHF Data Links

Future survivability considerations and the need for greater bandwidths point to the increased use of laser, millimeter, and submillimeter data links. Such links are inherently both secure and difficult to intercept because of their narrow beams, wide bandwidth capabilities, and the attenuation characteristics of the propagation medium. However, these factors also impose constraints on where, when, and over which paths they can be used.

8.9.17   Integrated Airborne CNI

Advances in microprocessors offer the potential for integrating some of the airborne communications, navigation, and identification (CNI) functions in the future. (This integration would, of course, have to be compared with the excellent "specialization" advantages that can be obtained when multiple processors are used.) The goal would be to make significant reductions in the life-cycle cost, weight, and volume of the airborne electronic equipment; additional capability for ECCM provisions might then be possible. JTIDS is one example of such a capability.

Two alternative steps toward integrating some airborne functions could be to add a digital data-link capability to JTIDS.

8.9.18   Fiber Optics

As discussed in Section 8.5, the military uses of fiber optics will be widespread in the future [29]. Some improvements could be obtained from fiber optics in the near term if an aggressive program were pursued now. The benefits of fiber optic use will be many:

- Weight, size, and potential system cost savings
- Elimination (or reduction) of lightning and EMP threats
- Negligible crosstalk, emissions, and pickup; this leads to added security and reduction of the ARM threat
- Relatively tamper-proof cables

With regard to capacity and bandwidth, 45 Mbits/s can be achieved today. In terms of the range-bandwidth product, more than 600 MHz km is possible today. Repeaterless link lengths of up to 8 km are now being achieved (signal attenuation of approximately 4 dB/km is being achieved regularly now). Because of all these factors, deployment flexibility is much better than with today's field cables.

Some of the remaining improvements sought for fiber optics are as follows:

- Connectors: multiconductor connectors that are easier to connect, are field maintainable, and have low losses
- Sources: improved lasers, drivers, and light-emitting diodes
- Detectors: increased sensitivity
- Resistance to "darkening" when exposed to nuclear radiation

8.9.19 Modern HF Networking

Modern HF technology opportunities (including adaptive HF techniques) indicate the possibility that HF may play an important role in tactical communications in the 1990s. If so, the system-oriented approach of taking advantage of the network aspects of HF links may provide further benefits. Because of the nature of the propagation medium and the locations of jammers, the best way to establish an HF connection from location A to location D may be to go from A to B to C to D, even though the overall distances may be an order of magnitude greater than the direct path. Adaptive routing and network management are key aspects of this approach.

The Navy, Air Force, and Army have been doing a considerable amount of research over the years in HF networking. HF communications is very useful for both long-haul (skywave) and short-haul (ground communications) for both tactical and strategic communications. Recently, HF communication has become more popular among the services in providing connectivity in critical battlefield situations. The Navy has used HF for years and is planning to use adaptive, spread-spectrum techniques in the future.

Described below is one of the concepts being considered by the Navy for an HF-Intra Task Force (HF-ITF) Network [30]. The HF-ITF Network is being designed as a robust, survivable, antijam (AJ) communication network for the interconnection of mobile task force elements. The HF-ITF Network must exhibit graceful degradation under severe stress. It must be designed for extended line of sight (ELOS) communication among a variable number of platforms (ships, aircraft, and submarines), be capable of handling voice and data, and satisfy specific precedence, timeliness, and error-rate requirements. It must also provide point-to-point and broadcast communication modes, and must interface with other networks in order to implement functional nets and/or serve as backup for long-haul traffic. The development of such a network is essential because fleet communication requirements have increased to the point where they far exceed the capabilities of existing communication systems.

The ITF Network will consist of various platforms (ships, aircraft) with markedly different characteristics, and is expected to support the traffic requirements of many diverse scenarios. The use of the HF (2 to 30 MHz) groundwave medium is dictated by its natural survivability properties in postnuclear detonation environments and by its ELOS communication range. At the same time, HF suffers from time-variable and unpredictable fading and self-interference, and as a result it is extremely difficult to obtain an accurate model for the HF channel. Consequently, the ITF Network must exhibit a high degree of robustness with respect to channel behavior uncertainties both in terms of its organization and in terms of the signaling used. Furthermore, equipment limitations place additional important constraints on the network.

In the design of such a network many choices must be made for highly interdependent variables. It is not prudent to make arbitrary choices for any of these variables to facilitate the determination of the remaining ones. Nor is it possible to determine simultaneously optimum or even simply "good" choices for all of them. One should proceed in a hierarchical fashion by choosing first those variables that either placed minimal constraints on others or allowed maximum flexibility in the subsequent choices. Three areas of design issues, although interdependent, could be pursued in a parallel fashion. These areas are:

1. The architectural organization of the network
2. The signaling issues (i.e., waveform and coding considerations)
3. The multiple access protocols for the sharing of communication resources

8.9.19.1 Design Choices

What are the design choices in the HF-ITF Network? They can be summarized best in the context of the Open Systems Interconnection (OSI) layering structure that has been proposed by the International Standards Organization (ISO) as a means of describing networking protocols, as discussed in Section 8.7.5. Layers 1 through 4 are of primary importance to the design of the HF-ITF Network.

At the first (physical) layer of the network, the function that must be performed at the transmitting node is the conversion of streams of digital symbols to HF waveforms, and at the receiving node the reverse operation must be performed. An important concern is the HF signal waveform design for broadband channels.

Waveforms are needed that provide rapid bit synchronization, efficient bandwidth usage, low mutual interference levels among users, and protection from jamming.

The second or link layer is concerned with transforming a raw transmission facility into a virtually error-free communication link. The highly variable quality of the HF channel implies the need for sophisticated error-handling techniques, coding, encryption, and interference rejection. Another important concern is the selection of protocols for controlling access to the communication link.

At the third or network layer, paths must be established between nodes that are not within communication range of each other. The establishment of paths requires the specification of the methods of switching and link management that include routing. Fundamental decisions include whether to use circuit or packet switching and whether to use centralized or distributed control.

The fourth or transport layer represents the interface between the command and control system and the communication system as well as the interface between different networks. The problems at this layer reflect back to problems in the lower layers as they concern operational "end-to-end" requirements such as response times and priorities, and they present new areas of concern such as user-to-user error and flow control.

The higher layers, namely the session (fifth), presentation (sixth), and application (seventh) layers, are not of direct concern to the network design at this time. They are user oriented and do not concern the transportation of information, but rather the users' terminal software for the establishment of handshake with the intended destination, the transformation of data formats, and the ultimate user application.

Evaluation of a complex system such as the HF-ITF Network is based on a hierarchy of performance measures. These are, in order of relative importance:

- Measures of survivability
- Measures of effectiveness
- Measures of efficiency

Measures of effectiveness refer to system performance primarily from a commander's viewpoint. These measures cannot be evaluated fully by analytical methods alone because of the complexity of the network. Thus, overall evaluation must also rely on simulation.

8.9.19.2 Network Architecture

The first major decision concerning the HF-ITF Network is the linked cluster organizational architecture (Figure 8.5). Many other design issues can easily be imbedded and studied in the framework of this architecture. Indeed, the desire to provide a framework in which other networking issues could be evaluated motivated early consideration of a network architecture. To ensure survivability, the proposed architecture is based on the use of distributed algorithms that enable

LEGEND

▰▱ INTER-CLUSTER LINKS
── INTRA-CLUSTER LINKS
□ CLUSTER HEAD
△ GATEWAY NODE
○ ORDINARY NODE

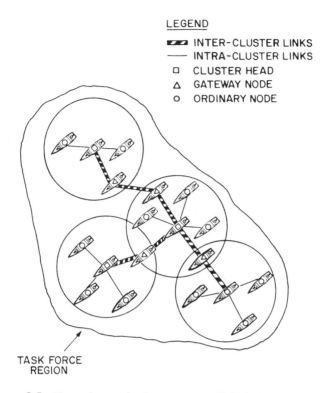

TASK FORCE
REGION

Figure 8.5   Network organization structure—linked cluster architecture.

task force platforms, a military unit consisting of a group of seagoing platforms
with a given fighting mission, to self-organize into a reliable network structure.
This structure consists of clusters of platforms within communication range of
local controllers known as cluster heads. The architectural profile of the network
at any given moment consists of clusters that either overlap or are linked to each
other via gateways. The set of links that interconnect the cluster heads and
gateways is known as the backbone network.

The linked cluster architecture is robust with respect to node and link losses.
It also strikes a compromise between efficiency in the use of the network's
resources and redundancy to ensure reliability. Thus, it provides alternate paths
for relaying and routing purposes, but keeps the number of these paths under
control. It is a flexible architecture that allows for many optional uses of the
links once they are established. Such flexibility is essential because a number
of constraints that could affect other design choices are not known precisely at
this time; these include equipment constraints, threats to network operation (including jamming and node destruction), and possible changes in communication
policy.

Under this architecture and its implementation algorithms, the network is
highly survivable, self-organizing, and automatically adaptive to link or node
losses. It permits the use of a variety of multiple-access switching and routing
protocols with either narrowband or wideband signaling. Finally, and perhaps

most important, the architecture takes full advantage of the apparent shortcoming and wide variation of the communication range achievable over the HF band. This has been accomplished by partitioning the HF band into a number of subbands, each with a bandwidth of a few MHz, over which the communication range for groundwaves is approximately constant. The organization algorithm is run consecutively for each subband, thus producing a set of overlaid connectivity maps that give rise to a set of simultaneously operating networks. The HF-ITF Network consists of this set of individual networks that are defined in separate subbands. Network management schemes will be developed to coordinate the operation of each of the individual networks into an effective overall ITF Network structure.

The length of time needed for the organization to be achieved in one subband is called an epoch (see Figure 8.6) and has been estimated to be on the order of a few seconds. If we consider M, say ten, such subbands, then the total time for the completion of one cycle of reorganization across the entire HF band is approximately one minute. This reorganization process can be executed continuously. Thus, once every minute and for a duration of about five or six seconds there would be an update of the connectivity map at one particular frequency subband (with a bandwidth of about 2 to 5 MHz). The rest of the time during each cycle the links corresponding to the previous connectivity map are open for normal communication, as indicated in Figure 8.6.

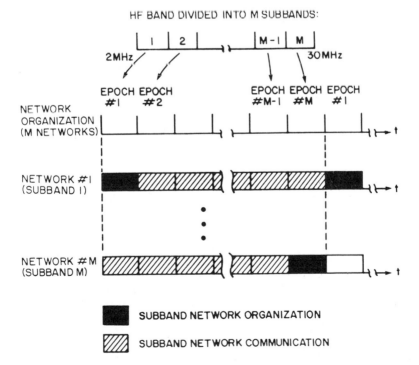

Figure 8.6 Timing structure for organization and communication under the linked cluster architecture.

8.9.19.3 Link Signaling Issues

The overriding factor in determining many of the ITF Network's features that distinguishes this network from other radio networks is in fact the special nature of the HF medium. The HF radio channel is both fading and dispersive, with propagation occurring via both groundwaves and skywaves. The ITF Network will rely primarily on the use of groundwaves to connect nodes, because although groundwave attenuation varies with frequency and sea state conditions, it is much more predictable and less dispersive than skywave propagation, which is supported at any time only over a portion of the HF band. Skywave signals must be considered as a source of multipath interference. The use of skywaves to supplement groundwave paths within the ITF Network and to link the ITF Network with other networks will be considered in the future.

In addition to multipath, sources of interference in the HF channel include atmospheric noise, local platform noise, and other-user interference. In the networking environment, other-user interference may be divided into two classes, i.e., interference resulting from platforms external to the network and interference caused by other network members (or other platforms with which the network is communicating). Interference caused by other task force platforms is characteristic of a networking environment, and it is clear that network management schemes are needed to minimize the effects of such interference. A major part of the HF ITF networking study has in fact been directed toward this problem. Interference arising from platforms external to the network and interference due to jamming cannot be controlled by the network. Since jamming represents the most significant limitation and threat to HF communication, it has been given special attention in the design of the ITF Network.

We have considered the signaling aspects of point-to-point communication and of multi-user communication for both narrowband and wideband environments. The following discussion of point-to-point communication brings forth clearly many of the issues. The subsequent discussion of multipoint communication complements this understanding by identifying additional issues that are unique to networking. This includes study of how narrowband signaling approaches will be replaced by wideband signaling as new equipment is developed, necessitating a hybrid form of operation in the interim.

8.9.19.4 Point-to-Point Communication

Narrowband signaling is not capable of providing satisfactory protection against anticipated jamming threats, even with the use of adaptive antenna arrays and selective band use. The use of wideband (spread-spectrum) signaling, however, can provide considerable protection from jamming. In addition, spread-spectrum signaling provides selective addressing capability, code division multiple access (CDMA) capability, inherent privacy and security, low interceptibility, and high resolution ranging. It has been concluded that among the candidate schemes for spreading in the HF channel, a pure frequency hopping (FH) system is the most practical choice for use in the ITF Network, because FH systems are more robust at HF than direct sequence (DS) or hybrid FH-DS systems. Specifically, FH systems are virtually immune to problems caused by differences in relative signal strength, are less sensitive to dispersion, and do not require a contiguous bandwidth. Also, the spreading factor can be considerably greater, and the pseudo noise (PN) code acquisition is easier to achieve and more difficult to disrupt for such systems.

Another issue in any form of signaling is the selection of the modulation method. Noncoherent M-ary frequency shift keying (MFSK) is a practical scheme for the HF-ITF Network because of its feasibility of implementation, its robustness with respect to fading and interference, and its compatibility with a wide range of hopping rates.

Bit error rate (BER) performance under worst case partial band (WCPB) noise jamming has been evaluated as a function of the equivalent signal-to-noise ratio for FH-MFSK signaling. Performance tradeoffs were developed for several alphabet sizes, many values of diversity, several coding schemes, both hard and soft decision receivers, known or unknown jammer state, and for nonfading as well as Rayleigh fading channels. From these results it is concluded that coding and/or diversity are essential for jamming resistance in the HF-ITF Network employing FH-MFSK. In fact, when operating under a high-quality data reception criterion (BER $= 10^{-5}$), AJ coding/diversity gains on the order of 35 to 37 dB are obtainable via the use of convolutional coding, with considerably reduced need for diversity. If the data reception criterion is relaxed to a medium-quality level (BER $= 10^{-3}$), the AJ coding/diversity gain is on the order of 16 to 18 dB.

The tradeoffs necessary to achieve acceptable levels of AJ performance have been analyzed in detail. For example, under some reasonable assumptions it is possible to operate links at a jammer-to-signal power ratio between 25 dB (at a data rate of 2400 bps) and 40 dB (at 75 bps) for a nominal case of 8-ary FSK signaling, a nonfading channel, 5 MHz spread bandwidth, 10^{-5} BER, soft decision receiver, worst case partial band noise jamming, and known jammer state. Results for 4-ary FSK are within 0.7 dB of those for 8-ary FSK. When the BER requirement is relaxed to 10^{-3}, and all other conditions are unchanged, the tolerable J/S ratio is increased by about 1.7 dB. Estimates of jammer-to-signal levels at which satisfactory communication can be maintained may have to be reduced somewhat after some issues related to practical implementation are examined.

8.9.19.5 Multipoint (Network) Communication

When one considers multipoint or network communication, the signaling issues become much more complex because of the potential interference among the users. In a narrowband architecture, there are a number of frequency slots to be managed ("switched") and distributed to the users. The new issue then is the development of link management schemes that permit the efficient apportionment of these frequency slots to the users via either dedicated or shared links. In a wideband architecture, such as that envisioned for the HF-ITF Network, the use of spread-spectrum signaling provides an inherent natural means of multiplexing the different users with acceptable levels of interference. This is achieved via code division multiple access (CDMA) techniques, in which, for the proposed choice of frequency hopping, each FH pattern corresponds to a distinct code. CDMA operation is usually asynchronous, and therefore it is possible for two or more users (using different hopping patterns) to transmit simultaneously at the same frequency, resulting in loss of data. The loss of data caused by such frequency "hits" can be handled via the use of coding and/or diversity.

The number of local users that can simultaneously use the same wideband channel is approximately one tenth of the number of frequency slots into which the channel is divided and over which the users are hopping. The exact number depends on factors such as signal-to-noise ratio, modulation scheme, coding, diversity, and acceptable BER criterion. Frequency reuse at distant parts of the network is of course possible. Nevertheless, the limitation in interference rejection capability that can be provided by a pure FH-CDMA system requires the development of additional control and channel-sharing ideas. Important issues specifically related to the use of FH-CDMA include synchronization requirements, hopping rates, the generation and distribution of FH patterns, and contention among signals that use either the same or different FH codes. Ideas for control and channel sharing come from the field of multiple access protocols (see Sections 8.7.1 and 8.7.3).

8.9.19.6 Multiple Access Protocols

While the linked cluster architecture provides a highly survivable organizational structure for the ITF Network, it does not specify how the HF band is to be shared by the task force platforms. What is needed is a set of multiple access protocols, imbedded within the linked cluster structure, that will permit the efficient allocation of bandwidth and equipment resources.

The choice of multiple access protocols for the ITF Network has been an important one since the initial stages of the network effort. The work on the architecture and on signaling has constantly focused attention on the access protocols that may be used in the network. The whole area of multiple access has received considerable attention in the last decade with the advent of packet switching in radio nets as well as the development of local ring networks. Traditional and straightforward schemes of distributing channel capacity by frequency or time division methods have been augmented by a variety of alternative strategies, each of which has its advantages as well as its limitations, making selection extremely difficult.

First of all, the nontraditional methods include the so-called code division approach, which accomplishes essentially what the frequency and time division approaches do, with improved AJ capability resulting from the use of spread-spectrum signaling. However, there are methods with radically different approaches such as random access or contention-based techniques as well as reservation schemes. These schemes were originally proposed and have been studied in the time domain. Currently under study are code domain versions of these schemes. Any of these methods may be used in a static (nonswitched) fashion or in a responsive, dynamic way that takes into account the status of traffic load and distribution in the network. Especially important is the compatibility of the multiple access protocol with the linked cluster architecture and with the selected signaling structures.

It is proposed to use dedicated links (actually fractions of links apportioned on a contention-free time-division basis) for much of the intercluster communication over the backbone network. Distributed channel allocation algorithms for this purpose are presently under development. Intracluster communication consists of communication between a cluster head and its members. Communication from a cluster head to its members could be implemented either on a broadcast or a point-to-point basis, depending on the type of traffic. The multiple access protocol used by cluster members to gain access to their heads will very likely combine elements of fixed assignment, reservation, and conflict resolution. A firm commitment to the details of multiple access protocol choices is not necessary until other design issues such as routing are also addressed. An important early decision, however, is the implementation of the access and switching methods partially in the code domain via the use of FH-CDMA. In all cases, owing to equipment limitations as well as limits on the number of spread-spectrum signals that can simultaneously be present in a local area, channel resources will also have to be shared in the time domain.

8.9.19.7 Baseline HF-ITF Network System Concept

To summarize, the work accomplished thus far in the design of the HF-ITF Network has resulted in a baseline concept for the architecture of the network, for the signaling methods, and for the multiple access protocols. This provides a comfortable framework for the study of the remaining design issues. Listed below are recommendations for a survivable HF-ITF Network design.

NETWORKING: Robust Network Design

- Number and Types of Nodes: 2 to 100; ships, aircraft, and submarines
- Task Force Dispersion: approximately 500-km diameter (ELOS communication ranges)
- Network Structure: overlay of several (expected to be between five and ten) networks in different portions of HF band
- Network Architecture: overlapping clusters of nodes within each of these several frequency subbands
- Network Control: hybrid (centralized local control within clusters, and distributed operation among clusters)

- Intracluster Operation: Protocols will include features of broadcasting, contention, and reservation. They must be compatible with frequency hopping spread-spectrum signaling.
- Intercluster Operation: Dedicated links over backbone network and auxiliary links; a distributed algorithm for link activation is presently under development.
- Topological Changes: adaptive and robust network control (continual updating of connectivities).

SIGNALING: Robust Waveform Design

- Frequency Band: HF (2 to 30 MHz)
- Propagation Medium: HF groundwave
- AJ Scheme: Frequency hopping spread spectrum
- Modulation: M-ary FSK (M = 4 or 8 are possible choices.)
- Receiver Detection: Noncoherent
- Coding and/or Diversity: Required to combat jamming, fading, and other-user interference. Both convolutional and block coding are being considered.
- Signaling-Related Parameters:
  —Data Rates: 75 b/s to 2400 b/s
  —Acceptable BER for data: 10^{-3} to 10^{-5}
  —Typical Hopping Bandwidth (bandwidth of one of the subbands): 2 to 5 MHz.
- Tolerable Received Jammer-to-Signal Power Ratios (Convolutional Coding and Optimum Diversity: 8-ary FSK, 5 MHz spread BW, BER = 10^{-5}, soft decision receiver, known jammer state, i.e., receiver can detect which symbols are jammed):

Worse Case Partial Band Nose
Jammed Nonfading Channel: $24.9 \text{ dB} \leq J/S \leq 39.9 \text{ dB}$

Rayleigh Fading Channel
with Broadband Jamming $22.7 \text{ dB} \leq J/S \leq 37.7 \text{ dB}$.
(Broadband noise jamming is worst case partial band noise jamming for the Rayleigh fading channel).

Future efforts will proceed within the framework provided by the baseline concept presented above. Several of the remaining design issues have been noted, e.g., the specification of multiple access protocols for intracluster communication and the continued development of link activation algorithms for distributed control. In addition, a number of network parameters still to be determined were noted, including the number of simultaneously operating networks (and related to it the bandwidth of each of these networks and therefore the frequency hopping bandwidth), and the alphabet size M for M-ary FSK signaling. A major signaling issue will be the acquisition and maintenance of synchronization of the frequency hopped signals. The tolerable J/S ratios presented above are based on an ideal receiver that uses soft decision decoding and is aware of which of the received chips have been jammed. These results can be easily extended to the case of a hard decision receiver, either with or without such jammer state information. Further work is needed to assess the impact of the nonideal nature of both the HF channel and the limitations imposed by practical equipment considerations.

8.9.20   Physical Counter-Countermeasures

An aspect of communications system planning usually left to other specialists is the use of physical counter-countermeasures such as ARMs. The use of ARMs against airborne and ground-based jammers could significantly improve network performance. Such physical counter-countermeasures, combined with direct aircraft and artillery attacks, would force airborne jammers to fly at greater distances from friendly territory and ground-based jammers to be more mobile and, therefore, smaller. In both cases, the effectiveness of the jammers would be reduced. The United States should improve its arsenal for anti-jammer ARMs.

8.10   JOINT-SERVICE COOPERATION

During the 1960s, the pressures of the Vietnam conflict required many rapid design and purchase decisions to be made concerning the tactical communications equipments for the U.S. services. Interservice interoperability did not have a high priority in the decision-making process. Consequently, much of the services' tactical communications equipment in the field today will not interoperate with "similar" equipment owned by the other services.

In the 1970s, a number of important actions were taken to increase the interoperability among the tactical communication equipment of the various services. Joint-service efforts such as TRI-TAC, JTIDS, and the ground mobile forces (GMF)/DSCS program will result in significant increases in interservice interoperability in the 1980s.

Nevertheless, many opportunities still exist for increasing joint-service cooperation in the 1990s, and each of these opportunities can result in improvements in the effectiveness of tactical communications equipment, especially in the areas of survivability and mobility. Some of these opportunities are discussed below.

8.10.1   Joint-Service Tactical Network

The Joint Multichannel Trunking and Switching System discussed earlier, if properly designed and operated, could provide significant improvements in the tactical ground-to-ground communications capabilities of the Air Force, Army, and USMC. Establishing many access points in a combat theater, with high-AJ links between these access/switching points, will provide all of the services with high-capacity, high-quality communications as the battle ebbs and flows. Ideally, as a service unit is relocated in the battlefield, it could rapidly establish a short-range high AJ link to the nearest joint-service access point. These access points would be manned by different services, depending upon the types of units in the area and the theater commander's plans. Decisions concerning which subscribers are given access and service would be made on the basis of the precedence levels of the subscribers, not on the basis of the service to which they belong. This approach, which parallels that used in the Defense Communications System, can be made to work in the tactical arena as it has in the DCS, and it holds significant promise for the 1990s.

8.10.2   JINTACCS Communications

The Joint Interoperability of Tactical Command and Control Systems Program is an important effort intended to increase interservice interoperability and thereby improve the overall capability of the U.S. tactical fighting forces. Work on the communications aspects of this JINTACCS problem was slow in starting and will still profit significantly from expansion and acceleration. Joint-service cooperation and compromises would significantly speed up the fielding of inter-operable JINTACCS-comparable capabilities.

For the longer term, the increasing use of automated C^2 systems by the services becomes the driving factor in achieving interoperability. Potentially, large amounts of information can be exchanged among, and processed by, the automated systems. Unfortunately, information exchange conventions such as message standards, protocols, and data links have been developed independently by the individual services. A promising approach toward eliminating the existing incompatibilities has evolved from a former Air Force effort, Interoperability Through Structured Message Exchange (ITSME). A language specification has been developed for structured information exchange between any two computers that are independent of the selected communication interface. The ITSME approach has features that permit a subset of the community of users lacking the latest software updates to continue to interoperate.

8.10.3   TACP Communications

The Tactical Air Control Party is an Air Force element that interfaces with Army units at various echelons and coordinates close-air-support aircraft to support Army troops in the forward area. Important changes in the TACP communications equipment are currently planned, including the use of advanced JTIDS. Designing the future TACP communications equipment to allow the use of the forward-area Army communications system (e.g., SINCGARS or TRI-TAC's mobile subscriber equipment) would provide a cost-effective approach to communicating with Air Force centers further to the rear. This would require joint-service cooperation, which is in keeping with the role of the TACP (and its communications) in providing Air Force support to the Army on a cooperative basis. (Another possible joint-service alternative would be the use of a Class III JTIDS terminal.)

8.10.4   Other Trends

- Artificial intelligence techniques that allow satellites to be more autonomous and independent of vulnerable ground stations (e.g., self-contained position keeping and altitude control, self-diagnosis, and repair).
- Integration of the design of ESM/EW and radar equipment with communications equipment to minimize interference, increase reliability, and maximize the sharing of expensive, high performance components.
- Advanced antenna designs that can conform to the shape of military vehicles and adapt the beam shape to maximize the gain in desired directions and null the antenna gain in undesired directions (e.g., direction of a jamming signal or a suspected interceptor).
- Distributed network communications stations that transmit low frequency (160–190 KHz) groundwave signals that hug the earth's surface and are relayed from station to station (They would be harder to intercept and less susceptible to the type of atmospheric disturbances encountered in warfare than conventional radio signals.)
- Real time adaptive system control techniques that can rapidly restore, reconstitute, and rearrange the transmission network in response to system outages, degradations, and traffic load imbalances.

8.11   REFERENCES

1. Compton, Jr., R.R. "An Adaptive Array in a Spread-Spectrum Communication System." *Proceedings of the IEEE*, (March 1978): 289–298.
2. Munson, R.E. "Conformal Microstrip Antennas and Microstrip Phased Arrays." *IEEE Transactions on Antennas and Propagation*, (January 1974): 74–78.
3. Metzger, L.S. "On-Board Satellite Signal Processing." Lincoln Laboratories Tech. Note 1978-2, Vol. 1, 31 January 1978. This report is available from the National Technical Information Service Doc. AD-A052768.
4. Hamsher, D.H., Ed. *Communications System Engineeering Handbook*. New York: McGraw-Hill, 1967.
5. *Data Decisions Communications Systems*, 1984, a four-volume updated reference service, Data Decisions, 20 Brace Road, Cherry Hill, N.J. 08034, (609) 429-7100.
6. Brick, D.B. and Hines, J.W. "The Impact of VHSIC on Air Force Signal Processing." In *Proceedings of the 13th Annual Asilomar Conference on Circuits, Systems, and Computers*, Pacific Grove, Calif., 5 November 1979.
7. *IEEE Transactions on Electronic Devices* (Special Issue on Very Large-Scale Integration), April 1979.
8. *Electronics*. Annual Technology Update Issue. New York, McGraw-Hill, 25 October 1979.

9. Sumney, L.W. "VLSI with a Vengeance." *IEEE Spectrum* (April 1980): 24–27.
10. *Proceedings of the IEEE* (Special Issue on Microprocessor Applications), February 1976.
11. *Proceedings of the IEEE* (Special Issue on Adaptive Arrays), February 1976.
12. *IEEE Transactions on Antennas and Propagation* (Special Issue on Adaptive Antennas), September 1976
13. Ricardi, L.J. "Communication Satellite Antennas." *Proceedings of the IEEE* (March 1977): 356–369.
14. Brick, D.B. "Air Force Tactical C³I Systems." In *Proceedings of the 42nd Military Operations Research Symposium*, 6 December 1978.
15. Cummings, W.C. et al. "Fundamental Performance Characteristics That Influence EHF MILSATCOM Systems." *IEEE Transactions on Communications* (October 1979): 1423–1435.
16. Rosen, P. "Military Satellite Communications Systems: Directions for Improvement." *Signal* (November–December 1979): 33–38.
17. *IEEE Transactions on Antennas and Propagation* (Special Issue on Conformal Arrays), January 1974.
18. Flanagan, J.L. et al. "Speech Coding." *IEEE Transactions on Communications* (April 1974): 710–737.
19. Diffie, W. and Hellman, M.E. "New Directions in Cryptography." *IEEE Transactions on Information Theory* (November 1976): 644–654.
20. *IEEE Communications Society Magazine* (Special Issue on Communications and Privacy), November 1978.
21. *IEEE Transactions on Electronic Devices* (Special Issue on Displays and LED's), August 1979.
22. Ristenbatt, M.P. *MILCOM-83 Proceedings*.
23. Roberts, L.G. "The Evolution of Packet Switching." *Proceedings of the IEEE* (Special Issue on Packet Communications Networks), (November 1978): 1307–1313.
24. Fossum, R.R. and Cerf, V.G. "Communications Challenges for the 80's." *Signal* (October 1979): 17–24.
25. Coviello, G.J. "Comparative Discussion of Circuit vs. Packet-Switched Voice." *IEEE Transactions on Communications* (August 1979): 1153–1160.
26. Kahn, R.E. et al. "Advances in Packet Radio Technology." *Proceedings of the IEEE* (November 1978): 1468–1496.
27. Sussman, S.M. "A Survivable Network of Ground Relays for Tactical Data Communications." *IEEE Transactions on Communications* (1978): 1616–1624.
28. Cook, C.E. "Optimum Deployment of Communications Relays in an Interference Environment." *IEEE Transactions on Communications* (1978): 1608–1615.
29. Gallowa, R. "U.S. and Canada: Initial Results Reported." In "Optical Systems: A Review," *IEEE Spectrum* (October 1979): 70–74.
30. McGregor, W. et al. "Preliminary Systems Concept for an HF-Intra Task Force Communications Network (HF-ITF)." NRL Report 8637, August 1983.

8.11.1 Further References

1. Author TBD, Digital Comms, *IEEE Communications Magazine*, 1983.
2. Bennet, W. "History of Spread Spectrum." *IEEE Transactions on Communications*, (January 1983).

3. Blackman, J.A. "Switched Communications for the Department of Defense." *IEEE Transactions on Communications* (July 1979): 1131–1137.

4. Cook, C.E., Ellersick, F.W., Milstein, L.B., and Schilling D.L. (editors). *IEEE Transactions on Communications* (Special Issue on Spread-Spectrum Communications). COM-30 (May 1982).

5. Costas, J. "Developments in Spread Spectrum." IEEE Sponsored Workshop, 1983.

6. Curry, T. "Defense Applications of LPI." *MILCOM-83 Proceedings*.

7. Dixon, R.C. *Spread Spectrum Systems*. New York: John Wiley, 1976.

8. Ellington, T. "DSCS III—Becoming an Operational System." Defense Communication Engineering Center (DCEC) (1980): 1499–1504.

9. Gabriel W. "Adaptive Antennas." *Proceedings of the IEEE* (1978).

10. Gutleber, F.S. and Diedrichsen, L. "TRI-TAC Considerations of Electronic Warfare." *Signal* (March 1978): 52–58.

11. Holmes, J. *Coherent Spread Spectrum*. New York: John Wiley, 1982.

12. Houston, S.W. "Tone and Noise Jamming Performance of a Spread Spectrum M-ary FSK and 2, 4-ary DPSK Waveforms." In *Proceedings of the National Aerospace Electronics Conference* (June 1975): 51–58.

13. Huth, G.K. "Optimization of Coded Spread Spectrum Systems Performance." *IEEE Transactions in Communications*, COM-25 (August 1977): 763–770.

14. Klass, P.J. "Meteor Trails May Aid Communications." *Aviation Week & Space Technology* (15 August 1983): 166–169.

15. Koga, K. et al. "On-Board Regenerative Repeater Applied to Digital Satellite Communications." *Proceedings of the IEEE* (March 1977): 401–410.

16. Krasner, N. "Optimal Detection of Digital Signals." *IEEE Transactions on Communications* (May 1982).

17. Krasner, N. "Signal Detection." *MILCOM-83 Proceedings*.

18. *IEEE Transactions on Communications* (Special Issue on Satellite Communications.) October 1979.

19. Milstein, L.B., Davidovici, S., and Schilling, D.L. "The Effect of Multiple-Tone Interfering Signals on a Direct Sequence Spread Spectrum Communication System." *IEEE Transactions on Communications*, COM-30 (March 1982): 436–446.

20. Nicholson, D. and Rakestraw, W. "On Signal Design for LPI Communications." *MILCOM-83 Proceedings*.

21. Oetting, J. "An Analysis of Meteor Burst Communication for Military Applications." *IEEE Transactions on Communications* (Special Issue on Military Communications, 1980): 1591–1601.

22. Oetting, J. "Example of LPI." *MILCOM-83 Proceedings*.

23. Pickholtz, R., Shilling, D., and Milstein, L. "Tutorial on Spread Spectrum." *IEEE Transactions on Communications* (May 1982).

24. Price, R. "History of Spread Spectrum." *IEEE Transactions on Communications* (January 1983).

25. Price, R. and Green "Communications in Muiltipath." *Proceedings of the IRE*, 1958.

26. *Proceedings of the IEEE* (Special Issue on Fault-Tolerant Digital Systems), October 1978.

27. *Proceedings of the IEEE* (Special Issue on Packet Communications Networks), November 1978.

28. *Proceedings of the 6th Annual International Conference on Fault-Tolerant Computing*, Pittsburgh, Pa., 1976.

29. Ristenbatt, M.P. "Performance of Spread Spectrum." *IEEE Transactions on Communications*, August 1977.

30. Ristenbatt, M.P. "Comparison of Three Approaches to ECCM for Air Force Tactical Voice." *MILCOM 1982 Conference Record*, Volume S, pp. 6.4-1–6.4-6, October 1982.

31. Schilling, D.L., Milstein, L.B., Pickholtz, R.L., and Brown, R. "Optimization of the Processing Gain of an M-ary Direct Sequence Spread Spectrum Communication System." *IEEE Transactions on Communications*, COM-28 (August 1980): 1389–1398.

32. Scholtz, R.A. "History of Spread Spectrum." *IEEE Transactions on Communications*, May 1982.

33. Scholtz, R.A. "More History of Spread Spectrum." *IEEE Transactions on Communications*, January 1983.

34. Scholtz, R.A. "Tutorial on Spread Spectrum." *IEEE Transactions on Communications*, August 1977.

35. Seashore, C.R., and Singh, F.R. "MM Wave Component Tradeoffs for Tactical Systems." *Microwave Journal* (June 1982): 41–72.

36. Simon, M.K., Omura, J.K., Scholtz, R.A., and Levitt, B.K. *Spread Spectrum Communications*, 3 volumes, Rockville, MD: Computer Science Press, 1985.

37. Stein, K.J. "Forward-Looking Technology Explored." *Aviation Week and Space Technology* (29 January 1979): 195–202.

38. Sugarman, R. "Superpower Computers." *IEEE Spectrum* (April 1980): 28–34.

39. Sumney, L.W. "VHSIC." *IEEE Spectrum* (December 1982).

40. Torrieri, D.J., "Principles of Military Communications Systems." Dedham, Mass.: Artech House, 1981.

41. Tyree, B.E., et al. "Ground Mobile Forces Tactical Satellite SHF Ground Terminals." In *Proceedings of New Techniques Seminar, Digital Microwave Transmission Systems*, Department of Electrical Engineering, Princeton University, Princeton, N.J., pp. 49–53, 27 February 1979.

42. Van der Meulen, E.C. "A Survey of Multi-Way Channels in Information Theory: 1961–1976." *IEEE Transactions on Information Theory* (January 1977): 1–37.

43. Viterbi, A.J. "Spread Spectrum Communications—Myths and Realities." *IEEE Communications Magazine* (May 1979): 11–18.

44. Wozencraft, J.M., and Jacobs, I.M. *Principles of Communication Engineering.*" New York: John Wiley, 1965.

Appendix A

JTIDS

The Joint Tactical Information Distribution System (JTIDS) is a key development/acquisition program for the military service tactical communications in the 1980s. Because of its importance, a separate appendix is included here to provide an in-depth discussion of the program. Only simple TDMA (Phase I of the JTIDS Program) is discussed. Readers interested in further details should consult [1].

JTIDS is a time-division multiple-access, secure, jam-resistant, digital information-distribution system with multiple-net and relative-navigation capabilities. The system capacity is sufficient to support the operations of widely distributed tactical command-and-control elements, aircraft, surface ships, submarines, and other elements that will be both sources and users of information. In any net, selected information can be broadcast and a user can select any information or subset desired by, and intended for, that user. Additional nets can be set up if needed. A secure, jam-protected digitized voice capability is available as a tradeoff for a number of data channels (on the basis of equivalent total bits/second required). The spread-spectrum mode lowers the probability of enemy interception of the signal.

The net architecture is nodeless. Units operating on the main net have universal connectivity to all other units on the net for communications and position-location purposes. The loss of any unit does not cause these functions to degrade. Any terminal can act as a relay; therefore, aircraft in the relay mode become a node only on a temporary basis.

A JTIDS net is defined by means of a given code sequence that determines the pseudonoise and frequency-hopping modulation. All users who have the code sequence share all the information broadcast by each member, or they can select only the subsets that are of use. A channel is a repetitive subset of the net, where the repetition is equal to the data rate of the users of the channel.

Figure A.1 provides a view of the JTIDS TDMA transmission organization. Net participants are assigned specific time slots on an information net in which to transmit information. Condensed within the brief message format is a series of data updates that communicate the information the participant has to contribute (e.g., position and status). Users may report at a different rate, depending on

TIME ⟶

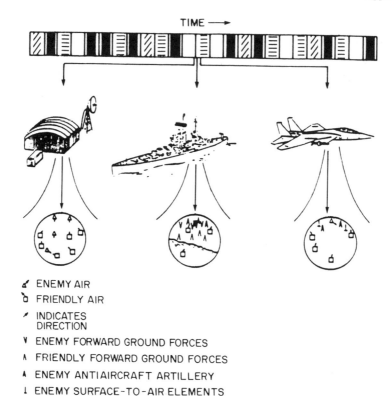

◁ ENEMY AIR

◖ FRIENDLY AIR

◢ INDICATES
  DIRECTION

ᴠ ENEMY FORWARD GROUND FORCES

ᴧ FRIENDLY FORWARD GROUND FORCES

ᴧ ENEMY ANTIAIRCRAFT ARTILLERY

⊥ ENEMY SURFACE-TO-AIR ELEMENTS

Figure A.1   JTIDS information filtering (artist's conception).

need, the quantity and criticality of the information they have and the nature of their roles.

Although the preceding paragraph examined JTIDS from the transmitter's viewpoint, JTIDS is basically oriented to the receiver's information needs, providing the receiver selectivity of information accessed. Because of the volume of information available (128,7.8125-ms time slots per second), the selectivity implies a means of filtering (Figure A.1) for information needed (e.g., by geographical area of interest, by information type, or even by data field).

A JTIDS epoch is defined as 12.8 min; this is the period of time after which time slots are renumbered (Figure A.2). Active participants must have at least one time slot in each epoch; passive participants can receive without being assigned time slots. One epoch contains 98,304 time slots, each 7.8125 ms long. Hence, if no information in the environment needs to be updated more often than once every 12.8 minutes, a single net's capacity would be about 98,000 participants, or 98,000 separate pieces of information per epoch. An intermediate time period, a frame, defined as 12 s, is of importance only in certain system timing operations.

The value of some information decays much more rapidly than every 12.8 mins or even every 12 s. Mobile elements change locations at speeds ranging from a walk to supersonic. The locations of air elements require a report at least every 30 s. A net can support 3,840 such reports, each updated every 30 s. But certain tactical elements must be assigned time slots considerably more often. For reports that require access in less than a second but are not sent too often, contention techniques are used to reduce the number of slot assignments required for the function.

The number of time slots assigned to each participant will be based on that participant's particular operational function, and on the amount of information the participant has to contribute to the net. Time slots can be assigned either singly or in a group in which the assigned slots recur periodically at a chosen rate.

The transmission within a time slot consists of a synchronization burst followed by a message. A guard time is included between successive slot transmissions to let the signal propagate throughout the line-of-sight environment. There are two possible guard times: one for normal range (300 nmi) and another for extended range (500 nmi). The various functions occurring during the time slot are discussed in [2]. The message length per slot is either 225 or 450 bits, with or without error detection and correction, respectively; the corresponding data

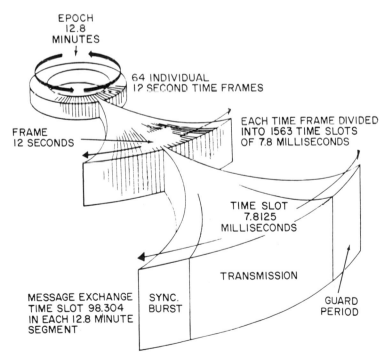

Figure A.2   JTIDS timing organization.

rates are 28.8 or 57.6 Mbits/s per net. Ahead of each message is a synchronization burst for receiver alerting and time-of-arrival measurement purposes.

JTIDS supports two basic message types: (1) formatted messages that are highly structured, i.e., each bit position has a specific meaning and, thus, a large amount of information can be condensed into a brief message; (2) unformatted messages that are essentially for any data that does not fit the standard JTIDS message formats, including digitized voice, ASCII messages, facsimile, any coded transmission, and high-speed data.

Relative navigation can be computed by a triangularization in JTIDS because all messages from other units represent direct radio range between units. Location with respect to a map or any other grid can be determined once the location of any member of the net is known.

Since security is an inherent characteristic of JTIDS, position and status reports transmitted from aircraft contain positive secure identification of the transmitting aircraft.

JTIDS operates in the 965-1215 MHz band. In its maximum jam-resistant and secure mode, the transmission pulses are spread and hopped over the frequency band by both pseudonoise coding and psuedorandom frequency-hopping techniques. Although JTIDS transmits in the TACAN band and across the IFF band, it has been demonstrated that interference with TACAN is negligible because of the wide bandwidth and low duty cycle of JTIDS. IFF interference is precluded by avoiding the specific IFF frequencies.

Appendix B

TRI-TAC PROGRAM

The TRI-TAC Program is currently developing equipment that will play an important role in the military services tactical ground-to-ground communications. This appendix summarizes the program and its equipment-development efforts; the reader interested in more details should consult [3] or the specific references cited below for the individual equipments.

TRI-TAC acquisition programs are identified in the main body of this book. The key programs are discussed under five categories:

- Transmission equipment
- End instruments
- Switches
- System control
- COMSEC/cryptoequipment

Transmission Equipment

The family of new troposcatter terminals currently being developed under the TRI-TAC Program is designated as the AN/TRC-170; it will comprise three configurations. All three configurations operate in the 4.4–5.0 GHZ band and have the following characteristics in common: secure, time-division-multiplexed digital transmission; a receiver noise figure of approximately 3 dB; a maximum data rate of approximately 2 Mbits/s; and transportability on standard military vehicles. The largest of the configuration employs two 15-ft parabolic antennas, two 6.6-kW transmitters, and quadruple diversity; its nominal service range is approximately 200 mi. (depending upon terrain, desired link availability, etc.). The equivalent parameters for the midsize configuration are two 9.5-ft antennas, two 1.85-kW transmitters, quadruple diversity, and 150 mi. Similarly, for the smallest configuration: one 9.5-ft antenna, one 2-kW transmitter, dual diversity, and 100 mi. Additional details concerning the TRC-170 can be found in [3] or [4].

With regard to line-of-sight links, the TRI-TAC Program plans include development of a new radio; the current designation for that radio is the Short-Range Wide Band Radio (SRWBR). It is to be used for line-of-sight transmission to a remote radio park, to replace cable for high-capacity interbase links, and to provide multichannel links for short, interbase trunks. It will probably operate at various rates up to approximately 20 Mbits/s at normal line-of-sight ranges. At the highest bit rates, the range may have to be reduced to maintain the requisite bit error rate.

End Instruments

This section discusses briefly each of these new TRI-TAC end instruments currently under development (more detailed information is available in [3]:

- Digital Secure Voice Terminal (DSVT)
- Digital Nonsecure Voice Terminal (DNVT)
- Advanced Narrowband Digital Voice Terminal (ANDVT)
- Modular Record Traffic Terminal (MRTT) equipment
- Tactical Digital Facsimile (TDF) equipment

The Digital Secure Voice Terminal (DSVT), also known as the Digital Subscriber Voice Terminal, is a four-wire, full-duplex, pushbutton telephone that provides secure access to the TRI-TAC circuit-switched network. Voice is digitized at 32 kbits/s (or 16 kbits/s) using a variant of delta modulation known as continuously variable slope delta (CVSD) modulation.

The Digital Nonsecure Voice Terminal (DNVT) is a four-wire full-duplex, pushbutton telephone that transmits and receives digital voice and loop signaling information at 32 kbits/s (or 16 kbits/s). It is compatible with the TRI-TAC equipment and can interoperate with a DSVT if the proper encryption equipment is provided at a TRI-TAC switch. The DNVT operation and voice digitization are similar to those of the DSVT, with the exception of voice encryption.

The Advanced Narrowband Digital Voice Terminal (ANDVT) will provide half-duplex encrypted voice communications capability at 2400 bits/s, using linear predictive coding; it will be compatible with "narrowband" channels (3–4 kHz). It will also be capable of operating in the data mode at bit rates between 300 and 2400 bits/s. Special interface equipment is needed to achieve interoperability between the ANDVT and the DSVT.

The Modular Record Traffic Terminal (MRTT) family of secure, ruggedized record-traffic equipment will provide facilities to compose, edit, process, transmit, receive, and distribute record traffic in a manner that is much more satisfactory than present-day teletypewriters. Two MRTT configurations have been identified thus far: a Single Subscriber Terminal (SST) and a Modular Tactical Communication Center (MTCC). The MTCC will be employed at facilities that handle a high volume of message traffic, the SST at facilities handling a lower message volume. The individual equipment will be compatible with other TRI-TAC equipment and also interoperable with some older equipment.

The Tactical Digital Facsimile (TDF) equipment will provide a wide range of facsimile capabilities for tactical use. These capabilities will range from a scanning resolution of 100×100 picture elements per inch to 200×200 picture elements per inch, from two shades of gray to 16 shades of gray, and from data rates of 1200 bits/s to 32 kbits/s. The TDF family will be able to transmit and receive photographs, fingerprint records, maps, map overlays, handwritten material, typewritten documents, etc.

Switches

The TRI-TAC Program will introduce several new transportable circuit switches and message switches. The automatic circuit switches being developed are designated the AN/TTC-39, AN/TTC-42, and SB-3865; the latter two, known as unit level switchboards, are significantly smaller than the TTC-39.

The TTC-39 will be modular, employing 120-termination analog-switching modules and 150-termination digital-switching modules. It will include a maximum of approximately 600 terminations, and the basic switch may be further expanded to a maximum of 2,400 terminations by collocating switches. Some of the TTC-39's service features, which include all those currently provided by the TTC-30, are conferencing (both progressive and preprogrammed), preemption capability (five levels), direct-access capability, automatic alternate rerouting, call transfer/forwarding, bypass capabilities for essential users, abbreviated dialing, and compressed dialing. The TTC-39 will handle voice, record traffic, data, facsimile, and imagery. See [40] for further information concerning the TTC-39.

The two unit level switchboards, the TTC-42 and the SB-3865, will be functionally similar except that the latter will be smaller and will have no internal COMSEC equipment. The TTC-42 will be available in 75- and 150-line versions. The basic SB-3865 has 30 lines, and the basic units can be cascaded into 60- and 90-line versions. Both switches will be capable of interfacing with the DSVT, DNVT, some older telephones, and other TRI-TAC switches.

TRI-TAC message-switching development is following a pattern similar to that for the circuit switches: a larger switch, the AN/TYC-39; and two smaller switches, the unit level message switches (AN/TYC-11 and AN/GYC-7). All of these switches will be automatic.

The TYC-39 will be a digital, secure, store-and-forward message switch; its basic functions will be to process, store, and account for and forward digital message traffic. The throughput can be up to 81 million characters per day, with a peak of 9,000 characters, and a maximum single-message length of 44,000 characters. It will have 25 (or 50) terminations, up to 25 (or 50) digital line termination units, up to 25 (or 50) quasi-analog modems, a digital transmission group capacity of 35 channels, an analog interswitch trunk-group capacity of 12 circuits; data rate capability for 45–16,000 baud, and up to 50 full-period dedicated circuits. The services that will be provided include six levels of precedence, eight security levels, data exchange between incompatible data terminals, message routing to other relay networks and switches, and automatic generation of some service messages. The TYC-39 will have both loop and trunk interface capability. Special cryptoequipment and inventory modems will allow terminals with analog modems and non-TRI-TAC cryptoequipment to access the switch.

The unit level message switches will provide local and trunk message switching for record and data traffic over digital and analog transmission links. Information bit rates can range from 75 bit/s to 16 kbits/s and transmission bit rates, up to 32 kbits/s. The AN/TYC-11 will be a 12/24-line shelterized message-switching central; the AN/GYC-7 will be a 12-line man-transportable switch to be used in forward areas.

The TYC-11 can serve as a concentrator for access to the TTC-39 or AUTODIN I. In addition, the TYC-11 can form independent subnetworks of unit level message switches and can serve dedicated/switched-access subscribers.

System Control

As currently envisioned, TRI-TAC's tactical communications control facilities will eventually include four hierarchical levels:

1. Communications Systems Planning Element (CSPE)
2. Communications System Control Element (CSCE)
3. Communications Nodal Control Element (CNCE)
4. Communications Equipment Support Element (CESE)

At present the functions to be performed by the CSPE have not been defined, functional specification for the CSCE has not been finalized, and Air Force plans for utilizing the CSCE and CSPE are uncertain. Current TRI-TAC plans call for the development of an initial version of the CSCE, called the Communications Network Control Center (CNCC), which will have limited capabilities.

As described in [3], the CSPE is a network's long-range planning and management facility. While it has not yet been designed, its primary functions will probably include estimation of long-range system requirements, contingency planning, determination of initial network configurations, allocation of resources, and DCS interface management.

The CSCE will provide dynamic, real-time control and management of a network or subnetwork and will allow traffic to be rerouted around destroyed nodes or jammed links. It will display information and provide data-processing aids to help the decision maker (network controller) select the most appropriate modifications to the network in the face of enemy-inflicted damage or jamming. Some of those modifications (e.g., changes to routing tables, trunk barring) will be able to be achieved rapidly while other actions (e.g., the establishment of new line-of-sight or troposcatter links) will take longer to effect.

A CNCE (which is subordinate to a CSCE) will be located at each node in an Air Force tactical communications network. It will provide the means by which communications resources at a node are assigned, monitored, and controlled. Along with the local switch, the CNCE will provide the interface between the users and the transmission facilities. Since the CNCE will operate initially in an environment consisting of a mixture of analog and digital equipment, it must interoperate with that equipment. The CNCE is designed on a modular basis so as to meet the requirements of various deployments and various mixes of analog and digital equipment. COMSEC equipment will also be provided to meet the communications security requirements of the CNCE.

A CESE will be part of most equipment developed by TRI-TAC; it will monitor its status and performance and send the resulting data to the CNCE. The detailed CESE design will be different for different equipment.

COMSEC/Cryptoequipment

TRI-TAC is developing a family of compatible COMSEC/cryptoequipment for tactical use. All members of this new family share a common, basic subsystem that may be used, in whole or in part, to provide secure voice. They are all compatible and span a wide range of security requirements, from the highest echelon to the lowest tactical unit. CVSD and audio-processing techniques provide analog to digital conversion and improved resynchronization.

In general, the TRI-TAC COMSEC equipment encrypts secure calls on an end-to-end basis. Before the information-exchange phase of a call starts, the parent circuit-switch sends a key to be used during the call to each digital securable end instrument. All trunk groups are bulk-encrypted.

In addition to the DSVT (described earlier), the COMSEC/cryptoequipment will include:

- a loop key generator
- an automatic key distribution center or rekeying control unit
- a key variable generator
- a trunk encryption device
- a dedicated loop encryption device

Other, auxiliary COMSEC/cryptodevices are also being developed for TRI-TAC.

Appendix C

STANDARD FREQUENCY NOMENCLATURE

Band Designation	Nominal Frequency Range	Specific Radiolocation (radar) bands based on ITU assignments for Region 2
HF	3–30 MHz	(Note 1)
VHF	30–300 MHz	138–144 MHz 216–225
UHF	300–3000 MHz (Note 2)	420–450 MHz (Note 3) 890–942 (Note 4)
L	1000–2000 MHz	1215–1400 MHz
S	2000–4000 MHz	2300–2500 MHz 2700–3700
C	4000–8000 MHz	5250–5925 MHz
X	8000–12,500 MHz	8500–10,680 MHz
K_u	12.5–18 GHz	13.4–14.0 GHz 15.7–17.7
K	18–26.5 GHz	24.05–24.25 GHz
K_a	26.5–40 GHz	33.4–36.0 GHz
Millimeter	40–300 GHz	(Note 5)

Notes:

1. There are no official ITU radiolocation bands at HF. So-called HF radars might operate anywhere from just above the Broadcast band (1.605 MHz) to 40 MHz or higher.

2. This is the official ITU designation for the Ultra High Frequency band. In radar practice, however, the upper limit is usually taken as 1000 MHz, L and S bands being used to describe the higher UHF region.

3. Sometimes called P band, but use is rare.

4. Sometimes included in L band.

5. No ITU radiolocation assignments. Experimental operation in this region has generally centered around the 94 GHz, 140 GHz and 230 GHz windows and at 70 GHz. The region from 300-3000 GHz is called the Submillimeter band.

APPENDIX REFERENCES

1. Eisenberg, R. L. "JTIDS System Overview." Principles Operational Aspects Precision Position Determination System, NATO Advisory Group Aerospace Research and Development. *AGARDograph* 245 (July 1979): 26-1-26-7.
2. "TRI-TAC." Joint Tactical Communications Office (TRI-TAC Office), Tinton Falls, New Jersey, September 1979.
3. Conner, W. J. "The AN/TRC-170—A New Digital Troposcatter Commmunication System. *Proceedings of New Techniques Seminar, Digital Microwave Transmission Systems*, Department of Electrical Engineering, Princeton University, Princeton, N.J., pp. 19–23, 27 February 1979.
4. Blackman, J. A. "Switched Communications for the Department of Defense." *IEEE Transactions on Communications* (July 1979): 1131–1137.

Glossary of Acronyms

A/C—Access Control
AABNCP—Advanced Airborne Command Post
AAT—Automatic Analog Testing
AC²MP—Army C² Master Plan
ACE—Allied Command Europe
ACOC—Advanced Communications Operations Center
ACTS—Advanced Communications Technology Satellite
ADA—Air Defense Artillery
ADDS—Automated Data Distribution System
ADP—Automated Data Processing
ADS—Army Data System
ADT—Automatic Digital Testing
AFSATCOM—Air Force Satellite Communications
AFSCF—Air Force Satellite Control Facility
AFSCN—Air Force Space Control Network
AGA—Ascension Ground Antenna
AI—Artificial Intelligence
AJ—Anti-Jamming
AJ LOS—Anti-Jam Line-of-Sight
AN—Analog/Nonsecure
ANDVT—Advanced Narrowband Digital Voice Terminal
ARM—Anti-Radiation Missile
AS—Analog/Secure
ASAC—All Sources Analysis Center
ASAS—All Sources Analysis System
ASW—Anti-Submarine Warfare
ATACS—Army Tactical Communication System
ATB—All Trunks Busy
ATDMA—Advanced Time Division Multiple Access
ATDS—Aircraft Tactical Data System
AU—Access Unit

AUNRI—Analog User Net Radio Interface
AUTODIN—Automatic Digital Network
AUTOSEVOCOM—Automatic Secure Voice Communications
AUTOVON—Automatic Voice Network
AWACS—Airborne Warning and Control System

BDE—Brigade
BER—Bit Error Rate
BITE—Built-In Test Equipment
BLOS—Beyond the Line-of-Sight
BN—Battalion
BPSK—Binary Phase Shift Keying
BSC—Binary Symmetrical Channel

c—Velocity of Light
C—Channel Capacity
C&D—Cover and Deception
C-E—Communications-Engineering
C/No—Carrier Power/Noise Density
C²—Command and Control
C³I—Command, Control, Communications Intelligence
CC—Command Center
CCITT—Comité Consultatif International Télégraphique et Téléphonique
CCS—Command and Control Segment
CDMA—Code Division Multiple Access
CENTAG—Central Army Group
CESE—Communications Equipment Support Element
CINCLANT FLT—Commander-in-Chief Atlantic Fleet
CINCNET—Commander-in-Chief Network

CINCPAC FLT—Commander-in-Chief Pacific Fleet

CINCUSNAVEUR—Commander-in-Chief U.S. Navy, Europe

CMD—Command

CNCC—Communications Network Control Center

CNCE—Communications Nodal Control Element

CNI—Communications, Navigation, and Identification

COMCEN—Communications Center

COMINT—Communications Intelligence

COMSEC—Communications Security

CONUS—Continental United States

COSCOM—Corps Command

CP—Command Post

CP/OPS—Command Post/Operations

CPCA—Camp Parks Communications Annex

CPE—Collocated Program Elements

CPI—Consumer Price Index

CRC—Control and Reporting Center

CRF—Channel Reassignment Function

CS—Communication Segment

CSCE—Communications System Control Element

CSDM—Continuous Slope Delta Modulation

CSOC—Consolidated Space Operations Center

CSPE—Communications System Planning Element

CUDIX—Common User Digital Information Exchange System

CVSD—Continuously Variable Slope Delta (Modulation)

DAMA—Demand Assignment Multiple Access

DASC—Direct Air Support Center

dB—Decibel $10 \log 10\ P_2/P_1$

dBm—Decibel relative to 1 milliwatt

DBS—Direct Broadcast Satellite

DCA—Defense Communications Agency

DCAOC—Defense Communications Agency Operations Center

DCS—Defense Communications System

DDN—Defense Data Network

DEWline—Defense Early Warning System

DFE—Decision Feedback Equalizer

DGA—Diego (Garcia) Ground Antenna

DGM—Digital Group Multiplexer

DID—Data Item Description

DISCOM—Divisional Command

DLED—Digital Link Entry Device

DLL—Delay-Lock Loop

DMSP—Defense Meteorological Satellite Program

DN—Digital/Nonsecure

DNA—Defense Nuclear Agency

DNVT—Digital Nonsecure Voice Terminal

DOCS—Defense Operational Control System

DoD—Department of Defense

DOMSAT—Domestic Satellite

DPCM—Differential Pulse Code Modulation

DPSK—Differential Phase Shift Keying

DS—Direct Sequence

D/S—Digital/Secure

DSA—Difference Set Array

DSB—Defense Science Board

DSCS—Defense Satellite Communications System

DSE—Direct Sequence Encoding

DSM—Data System Modernization

DSN—Defense Switched Network

DSP—Defense Satellite Program

DSVT—Digital Secure Voice Terminal

DTDMA—Distributed Time Division Multiple Access

DTMF—Dial Touch Multiplier Frequency

DTOC—Division Tactical Operations Center

DTS—Diplomatic Telecommunications System

E-3A—AWACS

E-O—Electro-Optic

EAC—Echelons Above Corps

EC—Earth Coverage

ECCM—Electronic Counter-
Countermeasures
ECM—Electronic Countermeasures
EHF—Extremely High Frequency
EIF—EUROCOM Interface Facility
EIRP—Effective Radiated Power
EJS—Enhanced JTIDS System
ELF—Extremely Low Frequency
ELINT—Electronic Intelligence
ELOS—Extended Line-of-Sight
EMP—Electromagnetic Pulse
EOD—End of Dial
ESD—Electronic Systems Division (Air
Force)
ESM—Electronic Support Measures
ESMC—Eastern Space and Missile
Center
ETS—European Telecommunications
Systems
EW—Electronic Warfare

FAC—Forward Air Controller
FCR—Flight Center Room
FDM—Frequency Division
Multiplexing
FDMA—Frequency Division Multiple
Access
FEBA—Forward Edge of Battle Area
FH—Frequency Hopping
FLOT—Forward Line of Troops
FLTSATCOM—Fleet Satellite
Communications
FLTTAC—Fleet Tactical
FS—Facilities Segment
FSK—Frequency Shift Keying
FTS—Federal Telecommunications
System

G/T—Gain/Temperature of an Antenna
GAPSAT—Gap Satellite
GB—Guard Band
GDA—Gimballed Dish Antenna
GENSER—General Services
GMF—Ground Mobile Forces
GPS—Global Positioning System
GSA—Government Services Agency
GSFC—Goddard Space Flight Center
GSTDN—Goddard Space Terrestrial
Digital Network

HAVE QUICK—Program to improve
aircraft radios
HF—High Frequency

I/O—Input/Output
IAC—International Access Prefix
IF—Intermediate Frequency
IFF—Identification—Friendly or Foe
IM—Intermodulation
INS—Indirect NICS Subscriber
INTACS—Integrated Tactical
Communications System
IOC—Initial Operational Capability
ISB—Independent Sideband
ISO—International Standards
Organizations
ITU—International Telecommunication
Union
IUS—Inertial Upper Stage
IVS—Internal Voice System
IVSN—Initial Voice Switched Network

JCS—Joint Chiefs of Staff
JINTACCS—Joint Interoperability of
Tactical Command and Control
Systems
JMTSS—Joint Multichannel Trunking
and Switching System
JSC—Johnson Space Center
JTIDS—Joint Tactical Information
Distribution System

KGA—Kwajalein Ground Antenna
KSC—Kennedy Space Center

LAN—Local Area ,Network
LCC—Launch Control Center
LDMS—Local Data Message System
LEASAT—Leased Satellite
LES—Leased Experimental Satellite
LF—Low Frequency
LOS—Line-of-Sight
LPC—Linear Predictive Coding
LPI—Low-Probability-of-Intercept
LSI—Large-Scale Integration
LWIR—Long Wave Infrared

M—Link Margin
M/C—Multichannel

$M = 2^n$—Number of passes in M-ary
  PSK (MPSK)
MARISAT—Maritime Mobile Satellite
MBA—Multiple Beam Antenna
MCC—Mobile Command Center
MCOS—Multichannel Objective
  System
MCS—Master Control Station
MDM—Multiplexer Demultiplexer
MEC—Minimum Essential
  Communications
MEECN—Minimum Essential
  Emergency Communications Network
MILCOM—Military Communications
  Conference
MILSATCOM—Military Satellite
  Communications
MILSTAR—Military SATCOM System
MRTT—Modular Record Traffic
  Terminal
MSC—Mobile Subscriber Central
MSE—Mobile Subscriber Equipment
MSG SW—Message Switch
MSK—Minimal Shift Keying
MST—Mobile Subscriber Terminal
MTCC—Modular Tactical
  Communications Controller
MUX—Multiplexer

N/C—Network Control
NAC—National Area Code
NASA—National Aeronautics and
  Space Administration
NASCOM—NASA Communications
NATO—North Atlantic Treaty
  Organization
NAVCAM—Naval Communications
  Area Master Station
NAVCOMPARS—Naval Computer
  Message Processing and Relay
  System
NAVCOMSTA—Naval
  Communications Station
NAVMAC—Navy Message Automated
  Communication
NAVSPASUR—Navy Space
  Surveillance
NBST—Narrow Band Secure Terminal
NCA—National Command Authority

NCF—Network Control Facility
NCS—Network Control Segment
NICS—NATO Integrated
  Communications System
NMCC—National Military Command
  Center
NMT—Nodal Mesh Terminal
NNT—Non-Nodal Terminal
NOC—Network Operations Center
NRI—Net Radio Interference
NSC—Navy Support Center
NSO—Normal Sustained Operation
NTDS—Naval Tactical Data System
NTS—Naval Telecommunications
  System

O&M—Operations and Maintenance
OCC—Operations Command Center
OCE—Operations Control Element
OSI—Open Systems Interconnection
OTCIXS—Officer in Tactical
  Command Information Exchange
  System
OTH B—Over the Horizon Radar

P^3I—Preplanned Product Improvement
PABX—Private Automatic Branch
  Exchange
PACCS—Pacific Air Command Control
  System
PADS—Packet Assembler
  Dissassembler
PBX—Private Branch Exchange
PCM Pulse Code Modulation
PLRS—Position Locating Reporting
  System
PLSS—Position Location Subsystem
PN—Pseudonoise
PO—Program Office
PSK—Phase Shift Keying
PSVP—Pilot Secure Voice Project
PT&T—Post Telephone and Telegraph

QPSK—Quadriphase Shift Keying

R&D—Research and Development
RAU—Radio Access Unit
RC—Resource Controller
RCC—Remote Command Center

RCOC—Regional Communications
  Operations Center
RCV—Receiver
RF—Radio Frequency
RFS—Request for Service
RLGM—Remote Loop Group
  Multiplexer
RMBA—Receive Multi-Beam Antenna
RPV—Remotely Piloted Vehicle
RTACS—Real-Time Adaptive Control
  System
RTS—Remote Tracking Station
RVCF—Remote Vehicle Checkout
  Facility

S/C—Single Channel
SAC—Strategic Air Command
SACDIN—Strategic Air Command
  Digital Network
SAM—Surface-to-Air Missile
SATCOM—Satellite Communications
SC—Systems Control
SCOTT—Single Channel Objective
  Tactical Terminal
SCPC—Single Channel per Carrier
SCPC-DAMA—Single Channel per
  Carrier Demand Assigned Multiple
  Access
SCT—Single Channel Transponder
SDI—Strategic Defense Initiative
SDS—Satellite Defense System
SEEK TALK—Past Air Force program
  to develop advanced aircraft radios
SGEMP—System-Generated
  Electromagnetic Pulse
SHF—Super High Frequency
SIE—Standard Interface Equipment
SIGINT—Signals Intelligence
SINCGARS—Single Channel Ground
  and Airborne Radio System
SLC—Submarine Laser
  Communications
SNR—Signal-to-Noise Power Ratio
  (usually in dB)
SOC—Satellite Operations Complex
SOPC—Shuttle Operations and
  Planning Complex
SPADOC—Space Defense Operations
  Center

SQPSK—Staggered (or offset) QPSK
  Modulation
SRWBR—Short-Range Wide-Band
  Radio
SS—Support Segment
SS-TDMA—Satellite-Switched Time
  Division Multiple Access
SSA—Solid State Amplifier
SSB—Single Sideband
SSBN—Strategic Ballistic Missile
  Nuclear Submarine
SSIXS—Submarine SATCOM
  Information Exchange System
SSMA—Spread Spectrum Multiple
  Access
SSS—Strategic Satellite System
SST—Single Subscriber Terminal
SSV—Space Shuttle Vehicle
STANAGS—Standard NATO Interfaces
STC—Satellite Test Center
STDN—Space Tracking and Data
  Network
SYSCON—Systems Control

TAC CP—Tactical Command Post
TACAMO—Take Charge and Move
  Out
TACAN—Tactical Air Navigations
TACC—Tactical Air Control Center
TACFIRE—Tactical Fire
TACINTEL—Tactical Intelligence
TACP—Tactical Air Control Party
TACS—Tactical Air Control System
TACSAT—Tactical Satellite
TAF—Tactical Air Forces
TAFIG—Tactical Air Forces
  Interoperability Group
TAFIIS—Tactical Air Forces Integrated
  Information System
TARE—Telegraph Automatic Relay
  Equipment
TASI—Time-Assigned Speech
  Interpolation
TCCF—Tactical Communications
  Control Facilities
TCE—Terminal Control Element
TCF—Technical Control Facility
TDC—Technical Data Center
TDF—Tactical Digital Facsimile

TDM—Time Division Multiplex

TDMA—Time-Division Multiple Access

TDRSS—Tracking and Data Relay Satellite System

TGM—Trunk Group Multiplexer

TH—Time Hopping

TMGS—Transportable Mobile Ground Stations

TOC—Tactical Operations Center

TRI-TAC—Joint Tactical Communications Program

TROPO/LOS—Tropospheric/Line-of-Sight

TS—Timing Subsystem

TT&C—Telemetry, Tracking, and Command

TTS—Thule Tracking Station

TW/AA—Tactical Warning/Attack Assessment

TWT—Traveling Wave Tube

TWTA—Traveling Wave Tube Amplifier

UHF—Ultra-High Frequency

ULCS—Unit Level Circuit Switch

ULMS—Unit Level Message Switch

USAREUR—United States Army, Europe

UTE—Unimpaired Tactical Effectiveness

VAFB—Vandenberg Air Force Base

VCO—Voltage Controlled Oscillator

VHF—Very High Frequency

VHSIC—Very High Speed Integrated Circuits

VLF—Very Low Frequency

VLSI—Very-Large-Scale Integration

VPD—Variable Power Divider

VRC—Vertical Redundancy Check

VS—Validation Segment

VSDM—Variable Slope Delta Modulation

VTS—Vandenberg Tracking Station

WB-CDM—Wide Band-Code Division Modulation

WHCA—White House Communications Agency

WSGT—White Sands Ground Terminal

WSMC—Western Space and Missile Center

WSU—Weather Support Unit

WWMCCS—World Wide Military Command and Control System

XMT—Transmitter

Index